Races of Men

Changes of the Human Race

By Steve Preston

2nd Edition

Table of Contents

Introduction

Have you been confused about how different races came about or who were the most ancient people? Are you one of many who believe that Noah started all the races? Do you believe all races came from Africa? Have you ever heard of the red or the blue race of people? So many questions need multiple answers. To try to get some reason in all this we will look at ancient texts, Haplotype studies, religion, and other sciences to build a reasonable history of race. That is about as good as we can get because so many seem to marry outside their ancestral lineages that it is difficult to determine where everyone came from exactly. One thing we will find is that the Cro-Magnon, Homo-Erectus, and Armenians help establish a timeline as do a number of ancient groups.

Biblical Reference

There are a number of "races" of people referenced in the Bible. Generally speaking these same races are affirmed in a number of ancient texts and physical evidence.

- The first was called the giants of old [Genesis 6] The Greeks called this group the Titans.

- The second was called the Anak people or the Nephilim. The Greeks revered this group ad the gods of Olympia. This race seems to have been red-skinned.

- The third was called the 6th day man that seems to correlate with what we now called the Homo Erectus. From many texts we find that this race was black.

- The 4th race was called the Adamic man, created on the 8th day or age. This race may be the beginning of the Armenian race or another that appears to be initiated near the area known as Canaan. This race seems to have been olive skinned.

There are many more, but I just wanted to get you going with this brief listing.

6th Day Man's Black Skin

There are several references to what this 6th day human was like and even his color is mentioned. The references from the Philippines, Africa, and Sumeria indicate that the 6th day man had black skin. This could mean that dark skinned individuals have at least some heritage associated with the very ancient 6th day man or it may mean nothing at all but it is information just the same. This does not mean in any way that black skinned people are less advanced; in fact if all things remained constant, they would be more advanced due to their longer heritage.

*Philippines-Kapre was one of the giants of old. He sits smoking a pipe waiting for people to walk past and cast spells on them. Most of the **Kapre are Black skinned**.*

Babylonian "Epic of Creation"- It was written about 1200 BC-*Blood to blood I join, blood to bone I join from an original thing, its name is MAN, aboriginal man is mine in making. They cut Kingu's [one of the Nephilim] arteries and from his blood they created man; and Ea imposed his servitude. Ea [God] had created man and man's burden. This thing was past comprehension. **Black-headed men will adore him on Earth**; the subjected shall remember their god. Let them serve the gods, work their lands, build their houses. Let*

black-headed men *serve the gods on Earth He created man a living thing to labor forever, and gods go free, to make, to break, to love, and to save. For your relief he made mankind, it is in the mouth of black-headed men who remember him.* [This is talking about the 6th day creation being black. Like all the other ancient texts, this creation was made to serve the Nephilim.] *The seed, created races of men from the world's quarters. From the stuff of a fallen god she made mankind.* [From the seed of the original 6th day man, other hybrid races were made around the world.]

Pleistocene Problem

The biggest problem in race determination is something called the Pleistocene Extinction. The Bible calls it Noah's worldwide flood, but what we know is that 10 thousand years ago, the earth shifted. Why all this happened doesn't help this investigation and there is way too much that would have to be explained, but what we find is that the shift destroyed just about everything. Massive flooding around the world, Mammoths were quick frozen as their pastures were now in the Arctic regions of Siberia. If mankind was a dumb stone Age sub human wandering around with a club and bare feet, he would have been killed along with most of the animals who died in massive piles at this time. Rather than simply saying races are impossible if they started after the Pleistocene event, we must understand that a number of people MUST have survived and these people carried DNA and features associated with their preflood societies. The Sumerians tell us that the Annunaki people huddled in the corner of their flying merkaba vehicles and cried out in fear as the world below them was being destroyed. Others have similar stories, but the only thing we need to recognize is that 10 thousand year ago, Noah landed on a mountain top and many others landed as well. The Bible says all that were on the ground died and there is massive evidence of the whole scale loss of life after the earth axis shifted. We can trace the Jews from Noah, but the other races will need some explaining. Before we can get to races after the great ending of the Pleistocene, we need to look at that time and even before. What we think we know about race might not be exactly what happened. The general accounting of race is provided next, but we will go much deeper. The chart below shows the timeline of some of the races to be discussed here. The first thing you will notice is that my timeline is more compressed than you are used to as the dates are in thousands of years ago rather than the millions you have been told previously. Another thing you see is that most of the races didn't make it past 10 thousand years ago when the Pleistocene Extinction killed off so many. There are also some unusual people. They need to be reviewed as well.

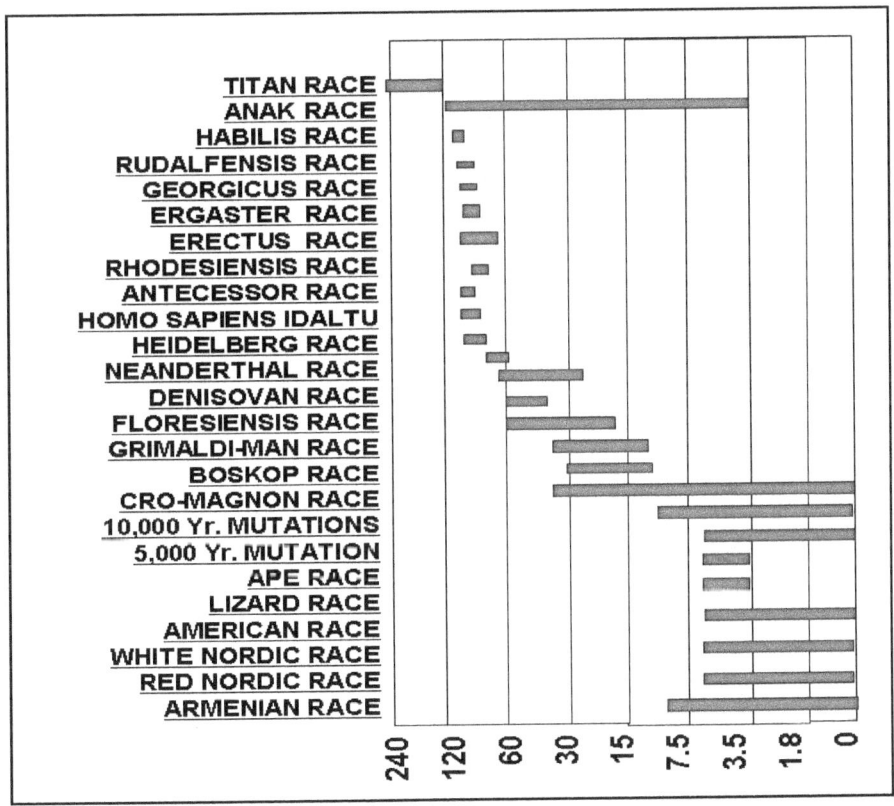

Race Generalizations

Let me just give you a less scientific accounting from before the great flooding. What we find is that races of people had already been established well before the end of the Pleistocene and one group called the Anak were rather strange. There is no more mystical group than the Anak, also known as Nephilim, Annunaki, or the Lord of Annu. Around the world this group survived in small settlements to reclaim control of the world as the water subsided. We do know some things by general characteristic, later we will add in the Haplotype DNA mutation testing to help clarify, but let's start with generalizations.

In the Americas they evidently were red skinned, high cheek boned, straight haired, light body haired, high forehead people. [Ancient texts identify region and color for us.]

In China they were yellow skinned, slant eyed, flatter featured, straight haired, light body haired, and round faced.

In Africa they were black skinned, narrow skulled, and dark curly haired.

In Northern Europe, they had the general characteristics of the Caucasian race.

In several records, there was also a bluish race that all but died off soon after the flood. We will look at those in a little bit.

Races Mix

Basic "races" began to mix together to form other races.

Red Skin

Various texts tell us that the "red skinned" humans went across to and established trade with the rest of the world. The trade umbrella included an outpost along the Mediterranean. They mixed with white skinned people and black skinned people. From their ancient writings, it is apparent that several cultural centers in Turkey, Phoenicia, and Egypt were trade outposts for a group we might call the Undal. [While I'm sure the area where the red race came from was not named Atlantis in the olden days, there is a substantial amount of evidence to give us assurance of the red race being located on the Island nation of Undal which include a certain Island named Atlantis by Plato and a number of other Greek, Egyptian, and Persian historians. One of the things we know is that in ancient times, the Phoenicians, Libyans, and Egyptians all called themselves redskins and many of the American Indians still have very reddish skin today.

Yellow Skin

We will test some of the conceptions of race later, but right now let's just think about what is believed on the surface. The yellow skinned people erupted on the scene about 6 thousand years ago and moved southeast to India, northeast to Mongolia and China, and west towards the Americas around 3500 years ago. Once they were in South America, they mixed with the red skinned people there and formed the pre-Inca civilization. Even though this may have been done, today we know that oriental people have substantially different chromosomes than those of the American Indian, Latino, black, and Caucasian groups. This means that the yellow guys were very isolated during much of this migration.

Black Skin

The "Black skinned" people are the oldest race and they can be traced back to the first Homo-Erectus and following evolutions. Generally speaking for almost 100 thousand years it appears the black skinned people did not move from Africa very often. About 11 thousand years ago, a small group moved up and east to India where they mixed with white skinned and yellow skinned people to form brown skinned people. About 6 thousand years ago, we find additional mutation to the "E' haplotypes and expansion and mixing with the others around Africa. Scientists today tell us that Latinos and black skinned people have many similar chromosomes and are more closely related than Caucasian, Oriental, or American Indian groups of people; as black skinned haplotype C people migrated to Spain and the Americas around 5 thousand years ago.

White Skin

The "white skinned" race seems to be a mixture of the Olive Cro-Magnon race and the Red-skin people. Most of the varieties of White skinned people came after the great mutation of 60 thousand years ago. They moved across Russia mixing with the yellow skinned people and around 4000 years ago they crossed to begin the Eskimo tribes. Before.

Olive Skin

Then something strange happened. Adam was "created". Some call him the first Cro-Magnon. He was a new "olive skinned, Adamic-human" different from the rest. Adam sort-of looked like the other men but the big difference was not seen by physical characteristics. He had a different "Spirit" life element. This new human immediately started mixing with the surrounding humans, but some remained as Jewish purebloods. Cro-Magnon people mixed with other groups are known as Gentiles in some religions.

Adam Was Not a Red Man

It is known that Adam means "from the red Earth" and people have mistakenly believed that it means Adam was one of the original "red men". The evidence suggests that the "red-men", or more correctly, the red hybrids generally controlled the world prior to Adam's time and up until it time of the Bharata War [about 5500 years ago]. The Red people that migrated to the Americas and Egypt, almost certainly were hybrid people. They were part Anak [red-skin], Part Cro-Magnon [olive skinned, and part "6th day man" [The human strain that initially erupted as Homo-Erectus and was manipulated and cross bred before the end of the Pleistocene].

As I stated before Cro-Magnon was different. According to many texts, he was created separately and things fit together if we look at Adam and his lineage as a separate branch of humans. The model used was possibly the "red-man" of the day, so the ancient texts indicated that he used "red dust" in his creation. It is a general belief that Cro-Magnon could not have been a straight-nosed, straight haired, tall, and almost no body haired, red-tinted, long headed person similar to the Anak people of the Pleistocene. Instead, he was medium sized [larger than Neanderthal and smaller than Anak], curly haired, had a larger nose, and had an olive-tinted skin color similar to the pure blood Jewish or Armenian races. The main factors for this conclusion are Biblical and Mutational [using Haplotype mutation tracking]. Mutation tracking shows the Jewish race did, in fact, begin in the Middle Eastern area and was significantly different that the original race of people that settled in Egypt and those who had been living in Africa at that time. To the south, the Adamic people mixed with the Anak, from Undal and in Egypt to form the beginnings of the Khemetians. They moved farther south and mixed with the black people to establish the Nubians. To the East they mixed with the Indians and formed the beginnings of the Sumerian nation, Scythian, and the Hittite nations. The Scythians moved east to make the Greek nation.

Let's See What We Have

Now that we have done all the mixing, we have Eskimos, red skins, and pre-Incas in the Americas. In Europe, we find white skins and Greeks. Egyptians are found in upper Africa and blacks are found in lower Africa. In Asia, we find all types including Indians, Jews, Sumerians, Hittites, Orientals, and Armenians.

Race Problem

Here is where I get into trouble. A portion of this race expansion didn't happen before the worldwide flood. In fact, we now know that about ½ the major mutations of man occurred about 11 thousand years ago and most of the remaining major changes occurred about 5500 years ago. We know that bones dug up from thousands of years ago show facial characteristics of similar people that died well before the flood time. "Ouch" Quit hitting me! I'm not trying to tell you that the Bible is a lie and that Noah didn't survive the flood. Quite the contrary; the Bible and science WILL agree and the Bible can still be established as TRUTH. What I'm telling you is that many race variations happened before the flood. There is no other logical explanation. I know that will take some explaining, but having them pop out a mere 10 thousand years ago, located in isolated regions around the world doesn't make sense. Somehow, the races survived the flood. Noah was not the only survivor. He and his immediate family were merely the only pure Cro-Magnon human survivors.

While some of you are getting angry that I talk about non-Noahites surviving the flood, I'm sure others can't even believe that races were initiated before the bad Pleistocene Extinction flood. Let's first see what the ancient historians said.

Worldwide Race Stories

Around the world, stories were told about how we became different. The stories show surprising similarities and unusual differences. They are presented here as a collage of colors. The stories are from the Easter Islands, the Hopi, the Cheyenne, Egyptians, and even from Urantia for completeness. Generally, they show inconsistency, but I think you may find some insight.

Easter Island Legend-*"First men to live on the island were survivors of the first race. <u>They were yellow</u>, very big with long arms. They had great stout chests, <u>huge ears</u> although the lobes were not stretched. They had pure yellow hair. Their bodies were hairless and shining. They came from the land behind America."*

Aztec Legend-*The world was initially divided into 5 regions ruled by 5 gods. A different color Black, white, blue and Red represented each of the four main godheads, Tezcatlipoca-Yaotl, and Quetzalcoatl. A different race of man came from each. <u>The **red "feathered god", Quetzalcoatl**</u>, lived in the east. The Line of Quetzalcoatl became the Aztec. **<u>The black wind god, Tlatlauqui</u>**, lived in the north. <u>The **blue plumed god, Huitzilopochtli**</u>, lived in the west. **<u>A different blue heron god, Tezcatlipoca-Yaotl</u>**, lived in the south. **<u>The white rain god, Taloc</u>**, lived in the center. He founded the city of T-Enoch-Titlan [City of Enoch].*

It seems evident that the colors represent actual skin color or it is unlikely that they would not have two gods with the same color. By the way, Quetzalcoatl means Precious feathered Serpent, so it appears that the Aztecs may not have been pure [or almost pure Cro-Magnon] humans. Having a serpent ancestor puts you in the hybrid class if you remember how Lilith and Samael turned into a serpent in the Biblical stories and handed Eve a fruit similar to a pomegranate which she immediately gave to Adam. [I know the Greeks changed the fruit to an Apple, but that has nothing to do with this history.]

Hopi Stone Tablets [Hopi Indians]-*After the Animal cycle, we went into the cycle of human being. They [the original gods] released the soul to humans. The humans lived on an island, which is now beneath the water. God said, "I'm going to send you to the four corners of the world." I'm going to change you to <u>four colors</u>. I'm going to provide you with "Original Teaching" in the form of two stone tablets." **<u>The red</u> people <u>were given guardianship of the ground</u>** and had to learn the cycles of time, healing, and farming. **<u>The yellow</u> people were <u>sent to the south</u>** and given the guardianship of the wind and had to learn about the sky. The Tibetans kept the tablets. **<u>The black</u> people were sent west** and given the guardianship of the water. Their tablets were put at the base of Mount Kenya. They will be the most humble. **<u>The white</u> people were sent north** and given the guardianship of fire. The Swiss kept the tablets. The keepers all live in the mountains.*

Cheyenne Traditions- *The Great Spirit created three kinds of men: red men, white men with hairy heads, and hairy men with hair all over their body. **<u>The hairy men</u> went to the <u>barren south</u>** and eventually dwindled in numbers and disappeared. **<u>The red men</u> went south after the Great Spirit <u>taught them culture</u>.** They went north again when the Great Medicine Man told them the <u>south would be flooded</u>. In the north, they found that **the white men** <u>had gone away. Another flood came and scattered them</u>, and they never came together again.*

Ancient Lower Egyptian Shilluk Tradition -*Humans were fashioned out of clay. Each region of the world had a different color clay. God created man from the available material. This made **some white, some red, some brown, and the Shilluk [black].***

Urantia Story-This came from "channeled information" so details should be treated as such. -*After the world was created, the most high God had his seven holy ones [archangels] create the animals through controlled evolution. Lucifer controlled Earth [called Urantia] and initiated 2 wars with heaven to enhance his control. Soon there evolved 6 different types of humans **about 100 thousand years ago**- The <u>Red race</u> controlled the Americas [the first race and most highly advanced], The <u>Black race</u> controlled Africa. [The last race and least advanced race] The combination of two races, a <u>Brown Race</u>, controlled India, The <u>Yellow race</u> controlled the Far East. The fifth race controlled upper Europe [color not described]. They became the <u>White race</u> when they mixed with the Adamic humans. God then made a higher level of human. Adam and Eve came to the Garden of Eden about 40 thousand years ago. They kept young by eating the fruit of the tree-of-life. <u>They were told not to mix with the other inhabitants.</u> Some of them did mix. Eve had sex with Serapotatia's main lieutenant [one of the early European race] and got pregnant with Cain. Afterward she had another son by Adam. [Serapotatia sure sounds like serpent to me.] Cain was born. After killing his half-brother, Cain left to the land of Nod, where he was revered because of his hybrid status. The <u>yellow race</u> came to South America about 25 thousand years ago and mixed with the red race as the pre-Inca society. A third war broke out under the command of Lucifer. The battles lasted 7 time periods. During the wars, the tree-of-life was protected by God's trusted Cherubim and Seraphim. After the wars, the entire Mesopotamian valley flooded and many were killed.*

There is a lot more about races in the huge Urantia book, but I just don't know about "channeled information. What is strange is the comment about 100 thousand years. For a long time science has relied on nuclear decay and sedimentation to build a historical timeline, but within the last 10 years, things have changed. Un-fossilized Tyrannosaurus and other dinosaurs less than 50 thousand years old, showed up, and testing objects near volcanoes showed to be hundreds of thousands of years old instead of months. Whenever sunspots showed high neutrino action on the earth, tens of thousands of years were sensed in a matter of days as the nuclear decay is now known to be almost useless. That being said, Ice core samples, paleo-magnetic tracking in the Atlantic, and Hot-spot movements in the Pacific Ocean showed consistency and has completely changed our timelines. For instance, the <u>end of the Cretaceous period that killed most of the dinosaurs was only</u> **120 thousand years ago**. I have no idea how the Urantia book, written in 1924, would not have any knowledge of these new findings. That brings us to Blue!

Blue People

We know about the "normal colors for people", but what about these strange blue ones? The bluish people discussed in some of the ancient texts may have been real and may have survived the flood. Five different instances have been recorded in recent history, which suggests that they may have survived in some small group. Rather than bluish skin, two of the reports are that the people had greenish skin as indicated below. At least two places indicated that bluish skin people were considered gods. Those places were Egypt and India.

India

In India, their highest god was depicted as being a blue being. Many artifacts attest to this fact. One is shown below to the right.

In fact, almost all gods of India are depicted as being blue. There is --

- *Venugopala- blue god of the musical flute*

- *Krishna the blue god of love*

- *Shiva the blue creator god*

- *Darshana the blue god of life*

- *Kali the blue death god*

These 'gods' are shown on the next left. I know it is in black and white, but take my word on this one.

These and others share one thing in common. They were all depicted as blue. They weren't blue because the Indians had too much blue paint. The coloration of these beings must have meant something. Someone had seen these beings in the past and had reported this feature to his descendants.

Egypt

In Egypt, some of the gods were clearly determined to be blue in color. The next image is a depiction of Osiris [on the right] and a normal [red] colored prince paying homage. Don't worry about Osiris' horns; just understand he has a BLUE body. Blue was certainly the color to be back then.

Ireland

Irish history is full of instances of greenish skinned people. At first, this seems fanciful because no green tint is evident in all known races, but the stories are numerous in quantity and may add credence to the claims.

England

Sometime between the years 1135 A.D. and 1154 A.D., two strange children were, apparently, found near Woolpit, England. The two children, a boy and a girl, were found terrified and huddled near a pit. They were screaming in an unknown language, and their clothes were made of a strange looking, unknown material, but stranger still, the report indicates that the children's skin was green. The children refused to eat or drink anything that was offered until someone brought in some fresh beanstalks. Soon after they were found, the boy sickened and died; but the girl became healthy and hearty, eventually losing the green hue to her skin. When she learned the local language, what she told of her origins only deepened the mystery. She said that she and her brother had come from a land with no sun; the people there, all green, lived in a perpetual twilight.

China

In China, they have found a huge quantity of terracotta statues of soldiers, and animals. It is almost like an army of lifelike beings. Many have seen these statues in various articles over the past 3 or 4 years but you might not have noticed one thing. Besides the large quantity of statues, the most unusual thing is that the statues have been made to have green skin. You would think yellow would be a reasonable color, but green was chosen. One possibility is that there were people with greenish tinted skin during the time that the statues were produced. [It would be kind of a mix between yellow people and the elusive blue people.]

Spain

In 1887, there was a curious Spanish story very similar to the story from England. By the testimony of many in the town, 2 children were found outside the city of Banjos, Spain. Like the English duo, they spoke an unintelligible language, but they could communicate with one another. They wore very unusual clothing, but unlike the English story, they had feature similar to the Negro race. The main thing here is that these children were also green in color. Similarly, the male child soon died, but the girl lived for a number of years and eventually learned some Spanish. She indicated that where she came from it was twilight all the time and across a great river, there was a land that had no sunlight. She and her brother heard a loud noise and they both found themselves in the field. She had no idea what the sound was or how they actually got where they were. There are so many similarities between the English one and this one that it is very possible, they are of the same encounter, but references seem to point to two separate events of green children.

Peru

Peru is where you can find blue people and I don't mean one of two. At the top of the Andes Mountains, researchers found an entire colony of bluish skinned people. It was believed that the bluish coloration was due to the thin air.

These people and the greenish tint people discussed above were possibly the last remaining evidence of the blue skinned people that were recorded in ancient history. The question might be is blue skin possible or is all of this whimsical. One answer might come from silver. Scientists know that a diet high in colloidal silver will turn your skin bluish and here is the neat part. The tint becomes permanent. Therefore, the coloration could have been associated with a diet rich in silver. As the silver content of food decreased, these people would begin to lose the bluish tint.

Red People, on the other hand must have lived for a much longer time and the remnants of that race are still around today. During ancient times, the Red Race was the most powerful for a long time. We can believe the Anak were Red and other grabbed onto the idea.

Red Skinned Ancestors

If you look at old artwork, Egyptian rulers were always painted with red skin. Various texts tell us that the "red skinned" humans went across to and established trade with the rest of the world. The trade umbrella included an outpost along the Mediterranean. They mixed with white skinned people and black skinned people. From their ancient writings, it is apparent that several cultural centers in Turkey, Phoenicia, and Egypt were trade outposts for the Atlanteans. While the area where the red race came from may not have been widely known as Atlantis in the olden days, there is a substantial amount of evidence to give us assurance of a large population of a red race came from a single location and it would have been a trade center. One of the things we know is that in ancient times, the Phoenicians, Libyans, and Egyptians all called themselves redskins and many of the American Indians still had very reddish skin today.

Red Rulers

If you understand just one thing after reading this book, remember this. The Anak went everywhere on earth. They settled and they took control. One reason we can believe their red skin is because the reddish skin of the American Indians, The Mayan description of Red rulers and general desire to be red around the world. Like the Germans of our later World Wars, the red Anak thought themselves to be a higher class that all the other people. Right now, I'm simply trying to show you something about the race of people that had the highest influence during a very long time in our history.

Desire to be Red

One way to understand the color of the most powerful people of a time is to see what people wanted to call themselves.

Phoenician-According to tradition, the Phoenician people came from an "Island of fire" located beyond the Indian Ocean. The Phoenician name means red man. Phoenicians are special in that the leader of the Anak race was identified as the King of Tyrus, the main city of the Phoenician lands. The Prince of Tyrus was also depicted as being one of the Anak. Ezekiel 28 provides detailed descriptions of both the one called Gadrael or Satan and his son. Both would have had reddish skin.

Egyptians- The Egyptians or Khemetians classified 4 groups of people similar to those identified in Indian Literature. The highest level was the "Rot" or red group and wrote about it on tomb walls. Almost all paintings of the early rulers depicted them RED. There gods and demi-gods also were depicted as being red skinned.

Arabs- The RED guys must have also ruled the various Arab groups. The "Himyaritic" Arabs were the "Red" Arabs. They were painted as being red on Egyptian monuments, so they were both honored and red.

Ethiopians- You guessed it Ethiopians jumped on the "want-to-be Red" bandwagon. Their name is taken from the word meaning sun burnt or red. Unfortunately, this seems to only have been a strong desire rather than fact.

Libyans- The ancient Libyans were called the Maxyans [interpreted as very red] by Herodotus. They would frequently go so far as painting their entire body red to appear more like their ancestors.

Algerians- The Algerian people called the Zuavas painted themselves red like the Libyans for a similar reason.

Americans- Many of the inhabitants of the Americas were and continue to have <u>reddish skin tone</u>.

PreMaya- <u>2 of the 4 gods of the Maya were declared to be red skinned</u> and stayed to make their civilization great.

The Hindu- The ancient Hindu depictions of all heroes, and <u>demi-gods were red skinned</u>.

I got too excited with blue and red people. Let me back up a little and start filling in some blanks. This first blank is a hard one to swallow, but after that things get easier as your viewpoint will expand. The first race of people was some have called the Homo-Gigantus.

Titan Race

This group is also known as **Homo Gigantus** by some, but you may have heard about them as the Titan people. As I mentioned, our Bible tells of 4 separate "developments of Man, but that is a tiny drop of information concerning races of mankind.

The first race described as ancient humans, Giants of Old, Archaics, or TITANS that walked with dinosaurs. They lived built cities, increased high technologies, and had wars during the Jurassic and Cretaceous period. Some believe that when they died, these Titans became the entities we call watchers or [angels] identified in Egyptian, Christian, and Jewish histories. During the Cretaceous period, many of the "angels" and Homo-Gigantus people initiated a massive war that destroyed the Earth. This war was thousands of years before Adam came along. It was the end of the Cretaceous Period and the world was left in shambled. Many know it as the Cretaceous Extinction. The books of Jeremiah and Isaiah in the Bible told us about the end of the war and the end of the civilization before the war that we can call the Titans. The book of Genesis simply called them the "Giants of Old".

Jeremiah 4:23-24*-[near the end of the wars] "I beheld the Earth, and, -- all the cities [Titan cities] thereof were broken down. **[If the cities were broken down, there must have been cities before the war.]***

Isaiah 9:17-21*---Is this the man [Satan leader of the Anak people] who made the world like a desert and overthrew its cities [Titan cities], All the kings [Titan Kings] of the nations lie in glory, each in his own tomb. **[Kings of nations and Cities shows a substantial civilization.]***

After the destruction, those responsible became a different type of "Human" [they mutated into what is commonly referred to as the Anak Race] and the controlled what was left of the world. A third race came along to help the ANAK people. This race was "created" during the 6th period or Tertiary Period after the end of the Cretaceous extinction according to the book of Genesis. We call this guy the Homo-Erectus. During the Tertiary Period Homo-Erectus changed quite a bit, but most references indicate that all animals were "manipulated" such that most were called "unclean abominations" during the Tertiary and into the Pleistocene Age. Sumerian, Chinese, Jewish, and many other historical references attest to some type of genetic manipulation that was not always successful, just like we do today.

Here is when the history gets strange as the Neanderthal Race, and many other races came from Homo-Erectus, but we are told they were not created by God. Instead, the ANAK tried to make Homo-Erectus a better servant, according to many texts. The outcomes were the Neanderthal, Heidelberg, Denisovan, Antecessor, Peking, and Rhodesiensis Races as well as others we will never know about. Today, we find that there are "alien genes" in Neanderthal man and in the Paracas Race, but most simply ignore the implications. A few of the hundreds of Paracas remains are shown next, including a small baby. We will look at them and similar people as we continue.

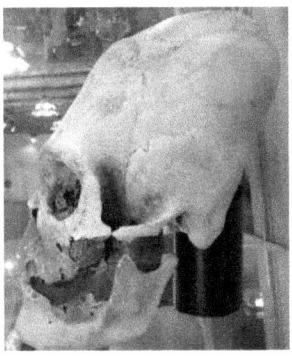 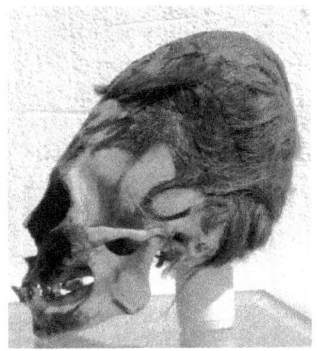 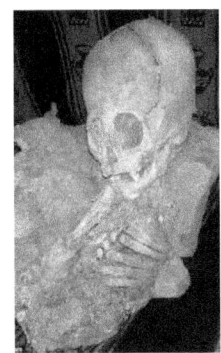

Finally, Cro-Magnon Man or Adamic man was created during the 8th Age which we typically call the beginning of the Pleistocene Age. 40 thousand years ago, As the Pleistocene Age ended, 10 thousand years ago, massive mutations caused substantial change in humans and many racial traits were formed. We know this type Haplotype DNA mutation tracking which has changed the way we view history. The Titan Race started it all. The next graphic shows some generalizations of this tracing of race.

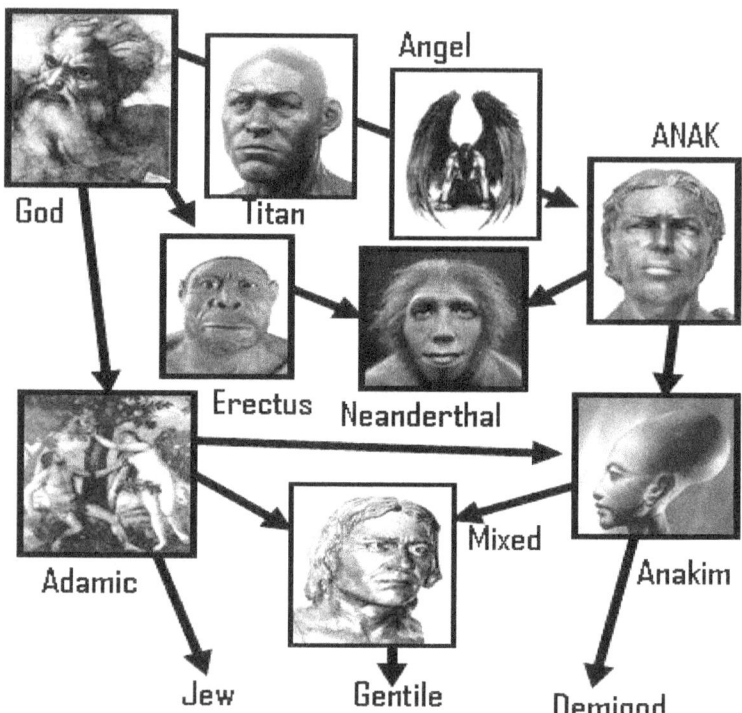

While not much is known about Titans simply because they were here so long ago, there has been physical evidence and written texts that help us understand about this first import race of man. Most of us have read about the Titans in the Greek histories, so I'm not going to get into that, but let's look at some of the finds around the world showing how this race of people lived around the world during the time of the dinosaurs.

Ecuador: Tools so ancient we cannot date them were found in Ecuador. The main thing that is noticed is that they could only be used by a massive race.

 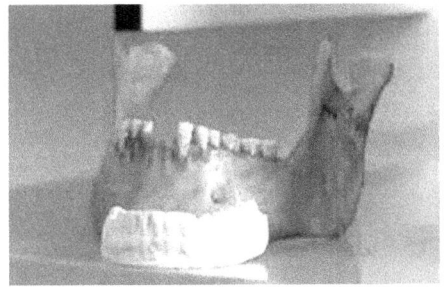

Around the World: Similar massive tools were found in Malta, Australia,

United States: While many might not know it, hundreds of giant skeletal remains have been uncovered in the United States. Here is the important part here. Some of these skeletons turned black when air hit them and they became powdered because they were so very ancient. One massive skull was found to be petrified. The size of the jaw was easily determined to be strange as shown above right. The "normal jaw is shown in front of the evidence of the Titan race in America.

Texas Feet: One thing is for sure if Titans lived here, they would have big feet and possibly even be found with Dinosaur feet; and so it is. A couple are shown below. The footprints are massive and show a race of 10 to 12 foot tall Titans lived during this time. Note the one of the right going the opposite direction as the Dinosaur as they shared the beach.

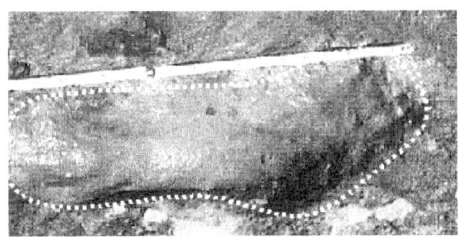 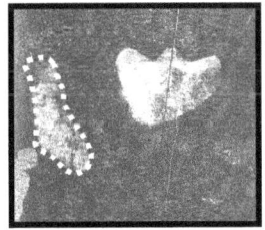

Found in Glen Rose Texas, this footprint is one of dozens of 18 inches long human footprints. Some of these footprints are found alongside dinosaur prints. The prints have been studies for years and conclusions force the reality that these giant

humans lived during the time of the giant dinosaurs. The image below compares a modern person foot with a cast of one of those from the Titans. Two of the racks shown are dinosaurs and the middle one is Homo-Gigantus.

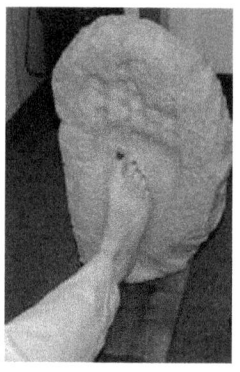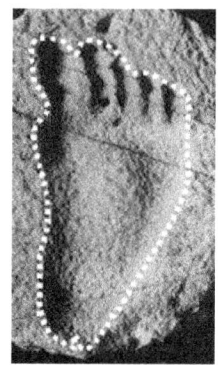

Many: I know you are thinking a couple of footprints can't be used to form an entire race of people, but what about hundreds? The more they look the more they find. Hundreds of human footprints alongside, dodging the dinosaurs are being found around the country. A couple of the many spots are shown below.

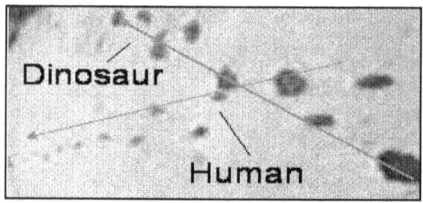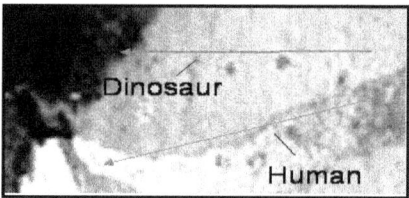

Dinosaurs & Giants-Texas is full of sites – At this one 14 human and 134 dinosaur tracks have been found together and estimated to be over 70 million years old.

Stone Bone: If these guys were here so long ago some of the skeletons would have been petrified. In 1982 a Researcher named Ed Conrad discovered in Pennsylvania petrified teeth inside the jaw-like area of solid rock. Then he found more and more specimens that bore the contour of human bone. You can just call it a stone head, but petrified skulls tell a tale.

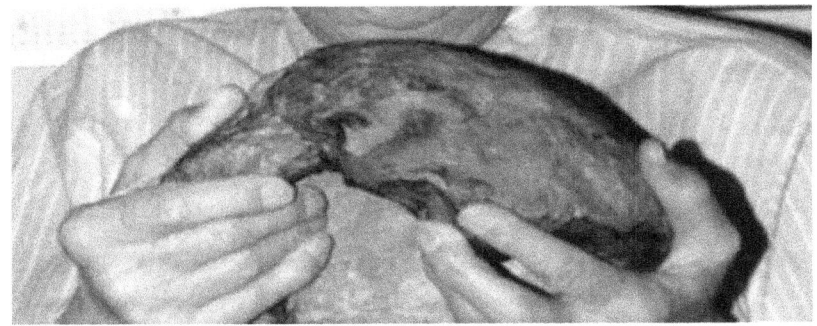

Wilton M. Krogman, the internationally acclaimed bone expert identified the first "stone bone" as a human calvarium, a portion of a skull with the eye-sockets broken off. A year later Ed Conrad discovered the large boulder in which was embedded the object that bore a distinct resemblance to a huge human cranium.

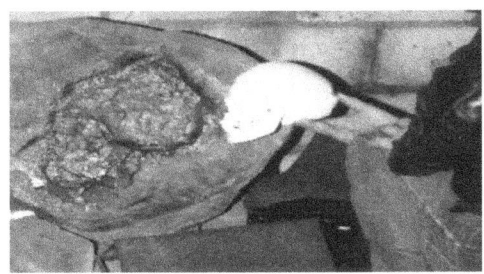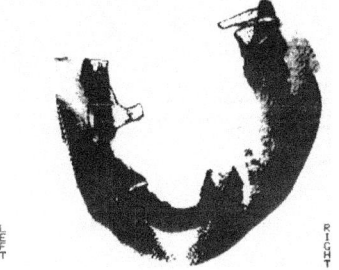

A CATscan had been done of this particular specimen and revealed intriguing characteristics of a human skull jaw and joints. [See above right]

Turkey: In the late 1950's during road construction in Homs southeast Turkey, many tombs of Giants were reportedly unearthed. These tombs were 4 meters long. During exhumation, the skeletal remains were examined. The human thigh bones were measured to be 47.24 inches in length. They calculated that the person who owned this Femur probably stood at fourteen to sixteen feet tall as shown to the right. A cast of this bone is shown below. Images of similar sized giants were shown in the United States, see below right.

  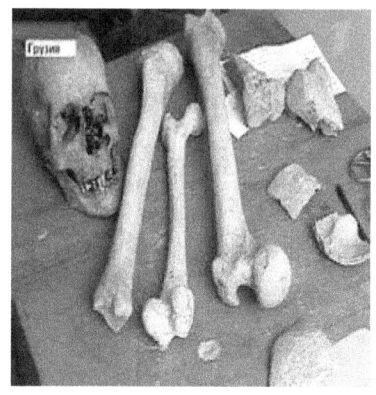

Russian Giants- Around the world we find these Homo-Gigantus bones. The picture above right shows a normal leg bone next to one of the giants found there.

Australian Titans: In Cosmo Newberry, In July 1970, an extensive, 4.5 kilometer long trail of giant-six toed footprints was found about 540 Kilometers north-west of Perth. Each print measured 38 cm long, displaying a soft pad and opposable big toe. The man's size was estimated to be about 11 feet tall. The researcher estimated that this human was only about an hour ahead of them. One of the footprints is shown next left.

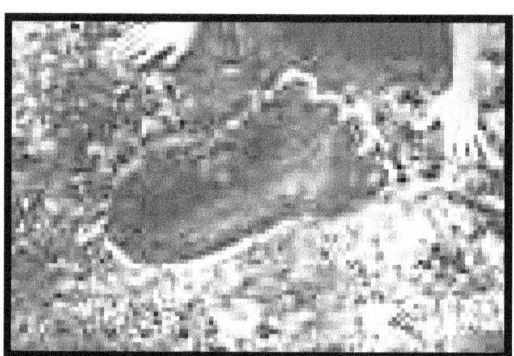 

Footprint Evidence: In sandstone beds on the Upper Macleay River, as shown to the right was found the prints of a monster.. One print was found with a toe 4 inches long and the total toe-span was 10 inches. The human would have been over 15 feet high. Near Kempsey, N.S.W. in sandstone beds on the Upper Macleay River. One print has a 5 inch long toe and the total toe-span is 10 inches - suggesting that the owner of the print may have been almost <u>17 feet tall.</u> One of the prints is shown to the right above. How would you like this guy to step on you?

Technology: Besides bones and feet, we have found a large amount of physical evidence of high social order and science of the Titans. Almost impossible to fathom, we have found shoeprints, batteries, nuclear power, brass manufactured good, and many other extremely ancient components of this once great society. After the war and the Cretaceous Extinction rocked the world, the Titans had vanished as did almost all the dinosaurs. On this barren Earth came a new race of men called the Anak.

ANAK Race

This race of men we can call the **Homo-Anakian**. I need to get into a little more details about the Anak race as they helped mold human races up until the massive war that caused huge numbers of mutations 5 thousand years ago. This race lived an incredibly long time [many thousands of years]. They were a major race of men from just after the end of the Cretaceous period 120 thousand years ago up until about 3500 years ago when the Biblical history and others tell us the last Anak or Anakim was killed by the great warrior David. I'm sure other people were identified as killing off the Anak in their regions of the world, but there was almost no life left in them so they would have been easy targets by that time. The Adena called them the Archaics and their history indicates they killed them all about 3500 years ago and we have found their mass grave in the Ohio valley.

Sorry for the 120 Thousand Years

I know you have been told the end of the Cretaceous period that destroyed the dinosaurs was 65 million years ago, but new evidence has confirmed the shortened time line after finding Nuclear decay to have drastically longer dates as effected by sunspots, heat, nuclear events, cosmic rays and all sorts of things giving us substantial errors in the past. For this book it really doesn't matter, but I think you need to see the big picture in all things that you do so please understand that the Comet that hit the Yucatan hit 120 thousand years, the event was recorded in the Arctic and Antarctic ice, in the magnetic fields at the bottom of the Atlantic and the Hotspot trail known as the Hawaiian Islands. The earth was completely destroyed. All the Cities of the Titans were gone and a new race of people became the rulers of this wasteland.

This group would modify DNA and genetically create animals all over the place. God would call those "manufactured animals" the "unclean or abominable animals". The Anak people also modified the new race of man we know as Homo-Erectus. Below is what the Greeks had to say. They reported there were 5 ages or creations of mankind, and the descriptions go along with the details presented here. This "multiple creations of man" isn't an isolated thought, but instead, was almost universal. The Book of Dzyan talks of 7 creations of man and many describe at least 4 creations. All attest to this **quasi-god ANAK** race of mankind.

Five Ages of the ANAK Race

Five ages of the ANAK were presented in Greek Mythology. Although not exactly creations, these periods go along with those identified around the world. This is after the time of the Premordial gods and the Titan Gods.

***The 1st Age*-**Cronos *[God] created the Golden Age. The happiest period, where people lived and died peacefully. There was no illness and no disease. The inhabitants never suffered from hardship of war or toil of the Earth. Foods were wild and plentiful. When they died they became spirits [ANGELS].* <u>*They became the guardians of mankind.*</u> *[This goes along with one of the probable histories. In it, the first creation of man, the titans, ended with men somehow being evolved into guardians of mankind or angels.]*

The 2nd Age*-When the* <u>*new gods [Anak humans] arrived, they*</u> <u>*began experimenting on the creation of mankind, creating a*</u> <u>*new Silver age.*</u> *Each succeeding age would be inferior from the last, from excellent to worse. The gods destroyed men,* *because they refused to honor them.* [The worked on the second creation of man and they essentially destroyed this human group [homo-Erectus] because they would not worship the ANAK men as gods. This second human is the one referenced in the Bible as being made on the 6th age of creation.]

The Third Age- *was the Bronze Age, which was populated with brazen men, who loved war for its own sake, until they destroyed themselves in* <u>*continuous warfare*</u>*.* [This is talking about the civil wars between various groups of ANAK after the creation of Adam.]

The Fourth Age-*This was followed by the Heroic Age. A race of demigods, heroes who would find themselves rewarded for their courage and heroic feats, at their death, in the Isles of the Blessed [heaven].* [A group of half Anak- half Neanderthals were called demi-gods. In this book I still refer to them as Anak people.]

The last age was the Iron Age. *This was the worse age, where* <u>*"good will" and decency would cease to exist. Men would*</u> <u>*suffer from great oppression by the wicked rulers. The rulers*</u> <u>*more than satisfied*</u> *their own needs. Because of their greed and thirst for power, Zeus would, finally, destroy this race.* [Things got so bad that god/Zeus mistakenly was exalted to this position, had enough. The end of the world was a huge flood. All of the wicked were destroyed. Only demigods and a few normal humans that worshiped God survived the worldwide flood. Note the similarity between Deucalion and Noah.]

I'm going to go through other examples very briefly, not because they are any less telling, but just to keep you focused on the ANAK.

DNA Starting Point

This whole thing about DNA and the beginnings of human life and all that have a problem when it comes to Homo-Erectus which is believed to have been the "Eve" of mankind out of Africa. The problem is that <u>there is NO DNA to study</u>. We simply have no DNA from Homo-Erectus, the Anak people or the Titans, so it is difficult to build up the structure from Homo-Erectus simply because they had similar features to us. A better starting point would be Anak, but there seems to be an issue that should be addressed here. The Anak may not have been able to procreate. The Bible indicated that there would be ENMITY between them and the descendants of Eve. The only way to affect the DNA of mankind was by manual manipulation of the genetic code. That being said, the Anak may have been able to procreate as the Greek indicated, but who knows!!! The Anak were considered gods.

Anak Gods Were Not Godly

One of the major problems with most historical references is that they try to make angels and the ANAK what they were not. They continuously try to make angels godlier than humans and the ANAK like gods. They have and had the same frailties, misconceptions, weaknesses, and egos that men have and had. The one major difference is that the angels were not bridled by carnal life. Even the ANAK humans knew much more than mere humans and had some spirit-being capabilities that could be used to get them in trouble. That is to say, the knowledge possessed by these beings and some fairly super-human strength made it much easier for these beings to get into trouble. No wonder people were confused. No wonder the belief in them and the God that made them is difficult for some. Here are some simple questions.

- *If they didn't exist, what did the people see, touch, learn from and tell about?*

- *How could they do the miraculous things that have been reported?*

- *How did they live so long? [Some of these humanlike beings may have lived over 40 thousand years according to multiple texts.]*

- *Why did everyone in the whole planet know about them?*

Don't go bringing up the weird gods identified previously as a reason to say that the gods existence were just made up. Let me give you an example. For a person to become a hero he has to do at least some heroic deeds. Surely, his exploits are expanded and you can typically find out about those exploits because different people make up different expansions. It is the kernel of the truth that emanates in the telling and the descriptions of the hero.

There is a "Big Kernel" in the stories that discuss the existence of gods.

Vain ANAK Gods

The Greeks may have been more correct than we would like to believe in their depictions of these beings. Following is a Graphic of Greek gods and goddesses. The thing to note is that these people were just like us and in many ways, worse.

Many ANAK were more vain, more lustful, and pettier than "Normal people" as below.

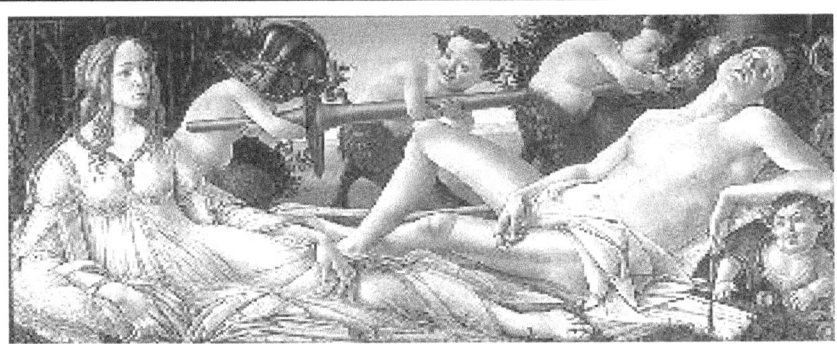

Preflood Anak Race

OK! Around the world, the almost identical story starts to emerge. Hopefully that wasn't tedious for you. The reason I brought out so many god items is to show you that the great similarities of these ancient texts truly confirm the accuracy of the histories. Certainly there are some differences as people tried to identify things they were not understanding completely, but the basic truths are described over and over and around the world. The truth is there were many races of giant rulers during the ancient times. The Bible and other ancient Jewish texts are also filled with many references to a race or races of giant humans that ruled over the smaller variety of human beings. I left these for last to sort of tie everything together. I think some of these texts give us a good general picture of life with the giant rulers, so I'm simply going to provide the verses directly. All the way through the Biblical history, the existence of giants was unquestioned, but somehow, today, we have a hard time believing that they ever existed. It's almost like a fear and the fear is so strong that evidence of these rulers is continuously suppressed in our school systems. Anyway, we are not going to suppress nor try to enhance possibility in this work. What will be provided is details gained from ancient historical texts, ancient religious books, ancient pictorial evidence, and actual physical evidence of this ruling class of humans. I'm not talking about people slightly taller than "normal" people. I'm talking about giants so huge that our Bible says "normal" people looked and felt as though they were grasshoppers.

According to the Bible, the last Anak or Anakim giants were killed by King David as the Israelites began taking over the modern world of the time. That time was about 1100 BC. Around the world, we will find that the huge rulers seemed to almost disappear about that same time. I know that doesn't seem reasonable right now, but it will.

With a "creation" time of 120 thousand years ago, these people were on the earth a long, long time. During that time, they mated or modified, Cro-Magnon people and primitive sub-humans and animals. Here are just a few of the many texts that reveal their genetic manipulative prowess.

"Book of Enoch" Anak

As far as ancient Jewish books that recorded the existence and problems associated with a race of giants, you cannot discount the book of Enoch. While it is an older book than most of the rest of the Biblical texts, it shows a side of these "other humans" that helps fill in the blanks. I've only picked out a few of the verses, but I think you can appreciate the writings regarding this subject. First, we will start with chapter 6. I'm going to provide several verses of this book because it paints a good picture of life before the worldwide flood with giant rulers controlling everyone.

Enoch 6-1-3 And it came to pass when the children of men had multiplied that in those days were born unto them beautiful and comely daughters. And the angels, the children of the heaven [Anak], saw and lusted after them, and said to one another: 'Come, let us choose us wives from among the children of men and beget us children.' [Essentially the same story is talking about modification of the human species, but ancient writers had little knowledge of what that meant.]

Enoch 7-1-3 And all the others together with them [ANAK] took unto themselves wives, and each chose for himself one, and they began to go in unto them and to defile themselves with them, and they taught them charms and enchantments, and the cutting of roots, and made them acquainted with plants. And they became pregnant, and they bare great giants, whose height was three thousand ells: [Not only did the Anak dabble in inappropriate sciences, but they taught people many secrets because they had existed for many thousands of years before normal sized humans came along.]

Enoch 7 Genetics of the Giants-6- And they began to sin against birds, and beasts, and reptiles, and fish, and to devour one another's flesh, and drink the blood. [There is no mistake here. Anak were modifying DNA.]

Enoch 69: 4-5 The name of the first [ANAK] Jeqon: the one who led astray the sons-of-God, and brought them down to the earth, and led them astray through the daughters of men. And the second was named Asbeel [another ANAK]: he imparted to the holy sons-of-God evil counsel, and led them *astray so that <u>they defiled their bodies with the daughters of</u>* <u>*men.*</u> *7-8 And the third was named Gadreel he it is who showed the children-of-men all the blows of death, and he led astray Eve, and <u>showed weapons of death</u> to the sons-of-men, the shield and the coat of mail, and the sword for battle, and all the <u>weapons of death</u> to the children of men.9 And the fourth was named Penemue [**Still another ANAK ruler**]: he <u>taught the children-of-men all the secrets of their wisdom.</u> And he instructed mankind in writing with ink and paper, 12-14 And the fifth was named Kasdeja [**the 5th of this group of significant ANAK leaders**]: this is he who showed the children of men all the wicked killing of spirits and demons, and the killing of the embryo in the womb, that it may pass away, and the killing of the soul the bites of the serpent, and the killing which befall through the noontide heat, the son of the serpent named Taba'et.* [This section of Enoch puts names to some of the "ANAK" that ruled over humans as gods. It describes inappropriate actions taken that eventually led to "discomfort" for all humans including the ANAK variety.]

"Jubilees" Anak

"Jubilees" also known as *"Apocalypse of Moses"* was an important work, and it gives a slightly different picture of the time of the giants, but it still confirms their existence. Fifteen copies of this book have been found showing its vast importance. It remained in the Greek version of the Bible for many years and in the Ethiopian Bible this book still remains as one of the major works, so it has pretty good credentials. Let's see what it said about our topic.

Jubilees 5:1-9 And it came to pass when the children of men began to multiply on the face of the earth and daughters were born unto them, that the angels of God saw them on a certain year of this jubilee, that they were beautiful to look upon; and they took themselves wives of all whom they chose, [This is a useful verse in that it calls the sons-of-God, the angels-of-

God, so there is less confusion.] *and they bare unto them sons and they were giants. And lawlessness increased on the earth and all flesh corrupted its way,-- and they began to devour each other,* **[There goes the eating again.]** *And God looked upon the earth, and behold it was corrupt, <u>and all flesh had corrupted its orders</u>, and all that were upon the earth had wrought all manner of evil before His eyes. And He said that He would <u>destroy man and all flesh upon the face of the earth</u>* [More reason to destroy everything with a flood.]

Jubilees 7:21-25 *For owing to these three things came the flood upon the earth, namely, owing to the fornication wherein the Watchers, against the law of their ordinances, went a whoring after the daughters of men, and took themselves wives of all which they chose: And they begat sons the <u>Naphidim,</u> and they were all unlike, and they devoured one another:* [Reassertion of the giant ANAK being born and the fact that they got fairly hungry from time to time.] *And the Giants slew the ANAK, and the ANAK slew the Eljo, and the Eljo mankind, and one man another.* [For this to make sense, I have to tell you about the Eljo. They were demigods half normal man, half ANAK. After the flood those who were first generation mixed humans were simply called Anakim. As this group intermarried with more "normal" people they were called GENTILES. It seems that during this 120 years, everyone was killing everyone.] *<u>And after this they sinned against the beasts and birds, and all that moves and walks on the earth:</u>* [The war was bad, but those who survived began reengineering the animals with genetic manipulation. If you have ever wondered why some animals were considered clean and others were considered "unclean" consider the unclean ones to be manufactured against God's instruction. This gives us another reason why the ANAK were considered "gods". They manipulated genetics to "sort-of" create new animals. The normal humans were in fear and threw them another virgin sacrifice, I'm sure.] *<u>And the Lord destroyed everything</u>*

from off the face of the earth; because of the wickedness of their deeds.

Please understand that this is a tiny listing of ancient texts that all say the same thing. The Anak somehow mated with Cro-Magnon and probably other Human-ish races either physically or by direct DNA manipulation and created new humans types and new animal types. Also it says there were massive wars and eventually many people and animals were destroyed even before the end of the Pleistocene.

Anak Physical Characteristics

For this we need to go down to Peru. What we find is many ancient skulls that don't fit into "normal" race characteristics. I think it is appropriate to consider these to be images of the Anak. I know normal historians say the people of Asia all went north towards Alaska crossing over to North America and ALL of them getting zapped into a different human with massive mutations. Then they travel for years without food until they finally get to the United States Areas, where they continue for centuries down to the tip of South America with only one major DNA change. If that isn't odd enough, they found hundreds of long headed skeletons and skulls along the southern portion of Peru. A few of the 300 found so far are shown below.

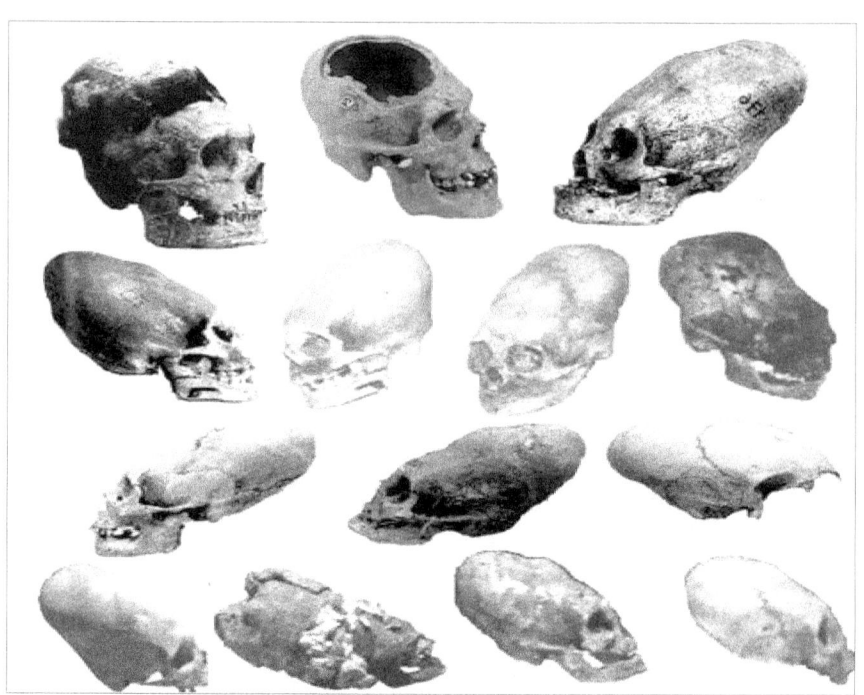

Paracas Elongated Skull DNA

The weird part was not that this was a new race of people; it was that the DNA had "Alien" genes similar to the Alien genes found in Neanderthal. In this case rather than being changed by a host, possibly, these are some of the Remains of the Anak or the half-breed ANAK and modern man we might call the Anakim. Recent DNA examinations performed on some of the hugely ancient skulls revealed shocking information that they might not have even come from known human beings. The study suggests they may have been a totally new human type. The skulls have some modified DNA which fails to match up with any identified genetic DNA material in GenBank. This GenBank is an open categorization database which contains every single piece of all recognized genetic records in the entire world. There was presented an indication that the DNA had mitochondrial DNA with various types of mutations that were totally unfamiliar in any primate, human or otherwise known to date. Some have tried to insist that the long heads were artificially made by squishing heads, but it is now known that the people were born this way.

Baby Long Head-Recently a small baby skeleton was found with the long head. Tests confirm it to be of the same grouping and the elongated skull was a baby with strawberry blonde hair. It is unlikely that the hair was bleached from age, or artificially dyed. The baby long head that I showed at the beginning is shown below left.

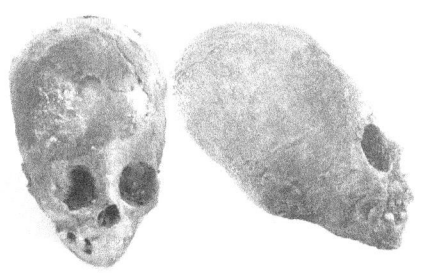

Remember the Redhead!-The picture below is another skull from one of these odd people showing the straight red hair similar to that found in Ireland. The Paracas, Peru skulls are different in a number of ways. Besides the long narrow head, the cranial volume is up to 25 percent larger and the skull is 60 percent heavier than conventional human skulls. Guess what red headed people usually have reddish skin. Please do not believe that the people that settled South America were walking travelers from Asia.

Russian Anak-The next group comes from Russia and the Anak also must have controlled that region for a time.

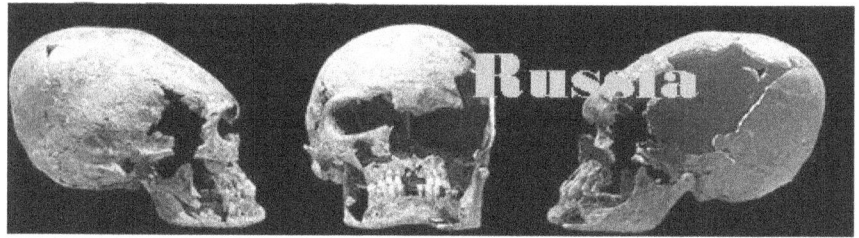

Other Anak Skulls-As many as a hundred of similar long skulled giants have been found in North American Graves to add to the list. In SW Asia, some more were found, Australia had some, and Ancient people in Cyprus had long heads. France elongated heads were found. The Czech Republic was another place. On and on we could go. The elongated head characteristic of the Anakim was worldwide. The following collage shows a number of them.

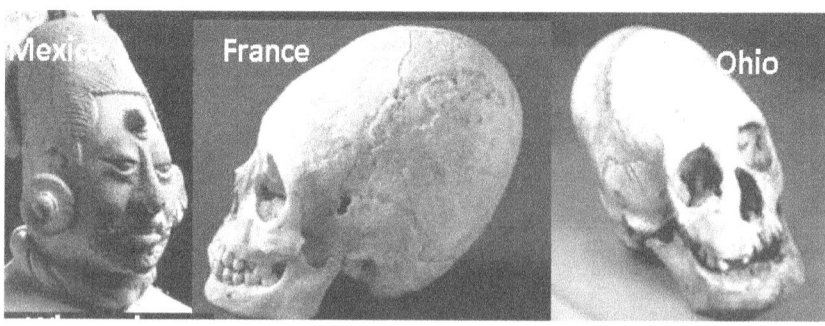

Scientists call these guys Dolicocephalic headed, but that's just because they like to say big words.

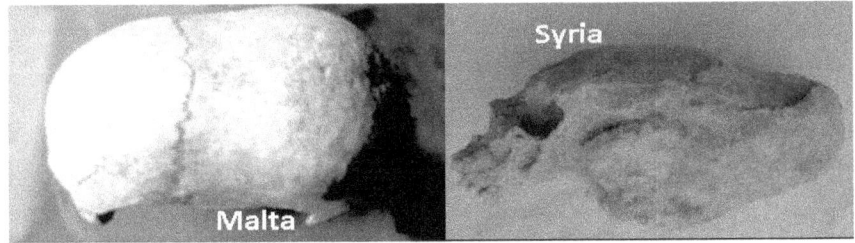

DNA Tests-The samples from 25 skulls consisting of hair, roots, a tooth, skull bone and skin and a geneticist in Texas began DNA testing. The results showed *"It had MtDNA (mitochondrial DNA) with <u>mutations unknown in any human, primate, or animal known so far</u>. But a few fragments I was able to sequence from this sample indicate that if these mutations will hold we are dealing with <u>a new human-like creature, very distant from Homo sapiens or Neanderthals.</u>"*

Aliens-One other thing should be brought up. Some sequencing has shown mutations that don't go anywhere. There are no known similar groupings. As done by many scientists, these "abnormalities" are simply disregarded and unimportant flukes. <u>Skulls from the preInca in Peru have some of these alien DNA and so do the Neanderthal</u>.

Traits of the Anak

Vanity: Really high; these guys really thought they were gods.

Size: They were not as large as the Titans, but much larger than "normal" people of that time. For those using Greek history as a baseline, these would have been the Olympians who thought themselves to be gods only to find out they later died.

Skull: Dolicocephalic to an extreme-Long thin skulls. Paddle boarding a baby's head to make it appear longer became very common around the world; to make people look like they were descendants of this group.

Teeth: Many ancient giant skeletons have been found with double rows of teeth, believed to be a trait of this group.

Fingers/Toes: A number of giant skeletons have been found with 6 fingers or toes on each appendage which is also believed to have been a trait.

Hair and Skin: They had reddish and straight to slightly curly hair. We can believe this race also had reddish skin coloration as many ancient cultures tried to claim red skin as their heritage to place them closer to the gods.

Civilization: They were highly civilized with scientific mastery of many things not understood since the Bharata War 5500 years ago and the massive mutations noted from that time.

When the Homo-Erectus Man came along, before the end of the Tertiary Period, the Anak went to work genetically modifying the races. Some of the humans did their own modifying by intermarriage and similar things, but the real mutations would occur about 11 thousand year ago just before the end of the Pleistocene. Before Homo-erectus came along, the Anak had the Homo-Habilis to contend with. We can believe the Anak scientists could not get much work out of Homo–Habilis. It had only a tiny brain, and feet much more suited for climbing. Genetic manipulation could not expand his capabilities to that needed for the Anak to had others to serve them so much of the work had to be done by them. Taking out trash, cutting down trees, farming etc. were menial and they wished for a new human to be created.

Genetic Manipulation

In this section we will look at one of the scientific achievements of the Anak called Genetic manipulation. I just read the other day that they are allowing human cloning as our geneticists go into some dangerous waters. The Anak were making all types of animals and modifying apes and ape-men until the great Homo-Erectus became a useful servant. We don't have to look far. Let me just give you a few to reinforce this important part of tracking the races of men. If you ever wondered how some popped up in the most unusual places, the answers have been in ancient texts all along. We can imagine that Homo-Habilis and even the Homo-Erectus had hair all over their bodies. Let's see what happens.

Indian History- *The first humans were covered with thick hair. When they mated [with Anak] they produced people as they are now.* [This doesn't mean there was sex involved, but we can imagine gene splicing was a popular science of the day.]

Sumerian Gilgamesh Story--*Aruru [God] pinched off some clay and created a [primitive man] Enkidu. His whole body was covered in hair. He knew neither people nor country, with cattle he quenched his thirst, a hunter and brigand—She [Shamhat –one of the Anak or Annunaki as the Sumerians described them] must take off her clothes and reveal her attractions. Do for the primitive man, as women [Annunaki women] do. She pulled not away, Enkidu was aroused.* [This is the most descriptive and shows that there were female Anak that had sex with male humans or artificially integrated DNA chromosomes, but the outcome was the same..] --*Afterward-the gazelles saw Enkidu and scattered, for Enkidu had* *stripped--- his body was too clean [the hair was all gone]. His legs were diminished-he could not run as before, he had become wiser.—Enkidu, you have become like a god [Anak]- He shall bring up daughters of gods [hybrid men].* [The Union between the Anak and the new human produced viable offspring.]

Mandaeans of Iran Story*-According to their traditions, the gods first made man. When he was finished, he looked like a man, but moved on all fours, had the face of an ape, and made noises like a sheep. Only later did he put in a soul and teach him and make him erect.* [This ape-man was not Adam.]

African Story*- In Africa, the same story was told. Hairy men became human after coming in contact with Anak.*

Southeast Asian Story*-An extremely hairy human "female" named Bota Ili was cooking food. A non-hairy fisherman named Wata Rian saw her and got her drunk. While she was asleep, he shaved her entire body. Only then did he find out she was a woman. She learned to wear clothes, they married, and they began a new race.*

Emerald Tablets [Egyptian Story]*-The master said-take them by the arts ye have learned of far across the waters until ye reach the land of the hairy barbarians, dwelling in caves of the desert. Follow there the plan.* [The plan was to inbreed with the hairy barbarians.]

Ngombe Tribe*-See if the Ngombe tribe folklore doesn't sound familiar. "A sky person [Anak] saw a hairy man [6th era man]. She married him and removed his hair. Then a Garden was made for man to live in."* [The offspring of this union was less animal like just like all the other descriptions.]

Inca legends*-The age of primitive man [hairy man?] was before the age of heroes [Anak and demigods] -*[Primitive man turned into heroes probably by a process called breeding.]

Aztec and Mayan history*-According to Codex "Laticano-Vatino" the first man [Homo Erectus man] was hairy. During the age of the four winds men turned into monkeys.* [This may be a reference to man with thick hair or something else we will look at later.]

Greek History*-The Greek Mythology was not as specific, but the Greeks did indicate that the early attempts at gods having offspring with man resulted in monsters. Here is what the Greeks said. "First only existed chaos and Gaia which would be the Earth was formed. Gaia gave birth to the sky. The gods [Ancient people] Gaia and Uranos had children-12 were titans, 3 were Cyclopes, and three were monsters with 100 hands. One of the titans took control [Cronos] and created the gods who took control by freeing the monsters. Zeus [leader of the Anak] had a woman created from clay [Pandora]. She open the box of knowledge and release evil into the world."* [Later Zeus became a lecher with human women as the Anak and humans mated. Zeus put out an edict that none of the Olympians were allowed to mate with humans, but that didn't seem to stop any of them.]

I know it sounds like the Anak were not modifying DNA responsibly like we are today. We know how very dangerous it is.

Responsible Genetics

Today scientists are much more responsible in what they do about making new animals. Most of us have heard about the puppies, fish, cats, and mice built to glow in the dark [1 and 2 in collage]. Others of the more responsible experiments include the fruit-fly grown to have legs coming out where antenna were [3], featherless chickens [4], big eared pigs and those with human organs [5], a human ear growing out the back of a mouse [6], gigantic animals like the monster bull [7], the sheep born with a human head [8], the mouse implanted with "created" memories [9], the dog with a second dog torso "grafted" to it.[10].

I just saw in the news England has granted it's OK to modify human genes to make a better baby, and we now have made an animal that is about 40% human to allow retrieval of body parts. While all of this seems well regulated and scientific and all, just imagine where the Anak went with it as we now know they even recreated entire Tyrannosaurus Rex, just like the Jurassic Park movies as we are finding remains less than 30 thousand years old.

The Bible even gave us a list of the more popular experiments. They were called "Unclean" animals that Satan and the other Anak had "miscreated". These included Eagles, Porpoise, Apes, all remade reptiles, and many more. Almost all the animals of the day were designed by these scientists. The *"Book of Giants"* found with the Dead Sea Scrolls provided an ominous warning indicated in the excerpt below.

For the Anak knew the secrets of heaven and sin was great in the Earth because of their experiments. They made mistakes and they killed many animals and people. They had sex with women and they begat giants. They selected two hundred donkeys, two hundred asses, two hundred rams of the flock, two hundred goats, two hundred other beasts of the field. The Anak performed unnatural acts, and begat giants and dragons. From every animal, and from every type of human was taken its seed for mixed sex. After a time they defiled the animals and people and begot giants, monsters, and dragons. God saw all that they begot, and, behold, all the Earth was corrupted with their blood and by the hand of man. --. The Anak thought that they would be saved and they would arise after death, but it was not so because they were lacking in true knowledge of heaven and because they were abominations of the Earth [Unclean].

The Anak had violated the laws of God by "miscreating". Some of the miscreated were considered human. If you remember from the Bible and similar ancient texts, God had compassion on the Anak, created animals, and had 2 of each enter the boat when Noah was saved at the end of the

Pleistocene. Around the world the same thing was happening. We are not up to that time yet. Possibly the first experiments were to modify the Australopithecine apes into a new race called Habilis.

Gigantopithecus Race

I'm sure the Anak worked on the Australopithecus as they began humanizing various animals. Don't even get me started with dolphins, but soon they just let their creative juices flow. With this experiment, they seemed to be trying to re-make the original Titan people. Scientists had this guy living between a million and 200 thousand years ago using the now debunked nuclear decay dating. Today's dating still places him in the mid Tertiary period, but that period evidently began only 120 thousand years ago so that places him about 90 and 80 thousand years ago. Gigantopithecus actually was manufactured during the time of Australopithecus and is put into that category by many. They found remains in what are now China, India, and Vietnam. The Gigantopithecus Race made up the largest ape-men that ever lived, standing up to 10 feet and weighing up to 1,200 pounds. I know you are thinking big foot but the Anak didn't care about scaring people. They were just having fun.

The first Gigantopithecus remains described by an anthropologist were found in 1935 by Ralph von Koenigswald in an apothecary shop. Fossilized teeth and bones are often ground into powder and used in some branches of Traditional Chinese medicine. This guy had huge teeth and bones so you could really make a lot of powder. In 1955 forty-seven Gigantopithecus teeth were found among a shipment of 'dragon bones' in China. I don't know why they didn't think they were real dragon bones, but that is not part of this story. From there more teeth and a rather complete large mandible was found and by 1958, three mandibles and more than 1,300 teeth had been recovered.

Graphics show a size comparison and some of the stuff they have recovered on the following page.

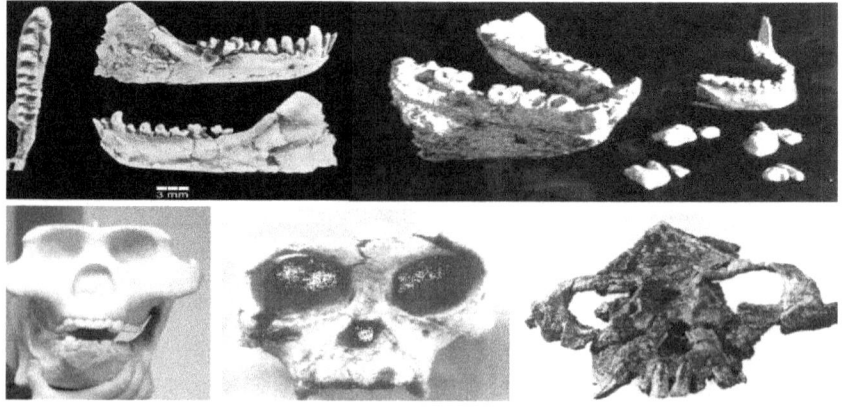

The tiny jawbone in the upper right corner of the previous collage is a human one. The following collage shows the large number of jaw and teeth finds.

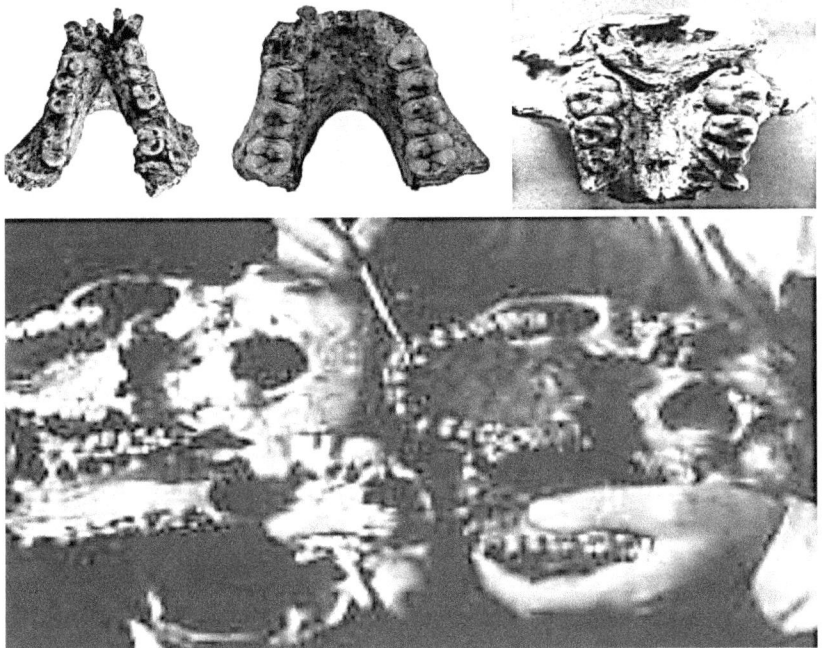

Meganthopus Race

After the gigantopithecus the geneticists decided to make something slightly smaller and the Meganthropus was created. This guy lived in the same general area as the previous model but he was more human looking and only 8 feet tall. As the skull shows, his teeth were <u>much more human</u> like, but the ridge along the brain was similar to that of today's apes. Certainly, the ridge is not as drastic as the gorilla, but this guy looked apish for sure.

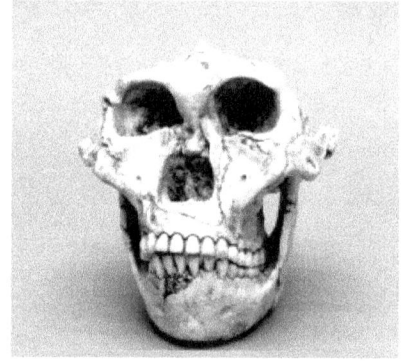

The first large jaw fragment was first found in 1941 by von Koenigswald. Unfortunately, Koenigswald was captured by the Japanese in World War II. Fortunately he managed to send a cast of the jaw to a scientist named Franz Weidenreich. Weidenreich described and named the specimen in 1945. It was the largest hominid jaw then known. The jaw was roughly the same size as a gorilla's, but was much thicker. The size was 2/3 the size of

Gigantopithecus, which was still twice as large as a gorilla. Meganthropus had a cranial capacity of up to 900cc and he roamed the earth about 90 to 85 thousand years ago, during the Tertiary period. He was here while gigantopithecus and the wide assortment of Australopithecines were making their mark. A couple of the skulls found were somewhat odd in that they had a double temporal ridge. That's the ape ridge on the skull of the large apes. I have no idea what that means, but it is interesting just the same. The geneticists decided to downsize and made the Homo-Habilis.

Habilis Race?

Homo-Habilis was not a true man in the strict sense. Homo Habilis lived during the Pleistocene approximately 2 to 1 million years ago by the old dating system [about 100 to 90 thousand years ago by modern techniques. While he was called HOMO- most place him Australopithecine. I'm just putting him here for completeness. He was an advanced ape.

Brain size: 600cm

Body: Habilis was short and had disproportionately long arms compared to modern humans.

Face: He had a less protruding face than australopithecines but had a cranial capacity slightly less than half of the size of modern humans.

Tools: He made and used primitive stone tools.

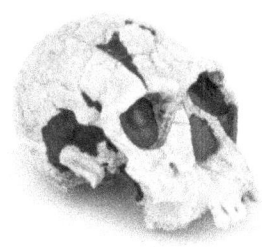
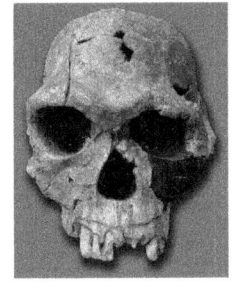

Hands: The Habilis didn't have an opposing thumb

Neck: He didn't have the powerful neck muscles, or the large occipital opening at the base of the skull, so he was much more comfortable walking monkey-like.

Thigh: His thighbones were curved like other apes.

Feet: His feet were still hand-like.

While having hands on your feet might be interesting, it is not a sign of the advanced "Humans".

Rudalfensis Race?

Like Homo-Habilis this was not a human. Skeletal remains from northern Kenya had been found: two jawbones with teeth and a face. The face was of a juvenile, but had features in common with the jaws. Homo Rudolfensis was born. Many believe this guy was also not human just like Homo Habilis.

Like many of the others in this grouping, Rudalfensis lived about 1.9 million years ago by nuclear decay or about 100 thousand years ago by newer methods. As Habilis and this guy lived about the same time, some have tried to determine a common ancestor, but to date these guys seems to just have sprung up around the same time.

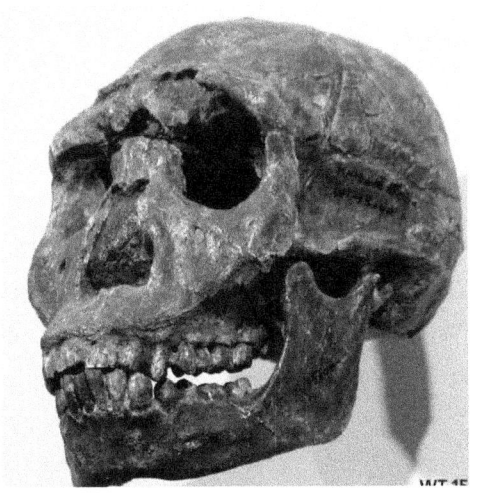
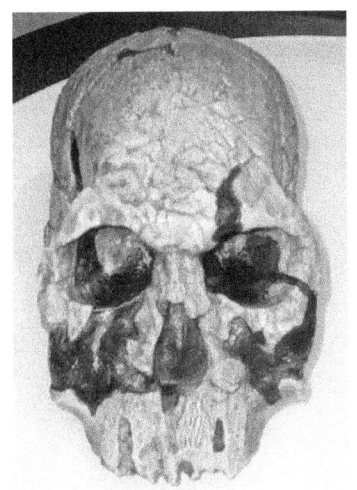

Skull: The construction looked very ape-like, possibly due to the large brow ridge.

Brain: The cranial capacity was about 700 cm³

Face: The face was said to be "incredibly flat", with a straight line from the eye socket to the incisor tooth.

Jaw: The jawbones were shorter and more rectangular than known Habilis. [See image above left]

Hands: The Rudalfensis didn't have an opposing thumb.

Neck: He didn't have the powerful neck muscles, or the large occipital opening at the base of the skull.

Feet: While we don't have much to go on, it is believed his feet were still hand-like and his thighbones were curved like other apes and the Habilis.

Georgicus Race

Out of Africa came a little faster than they originally thought. As Homo Georgicus was found in 2002 in Dmanisi, Georgia. This guy seems to be in between Homo Habilis and Homo Erectus. Possibly he was humanish.

A partial skeleton was discovered in 2001. The fossils are about 1.8 million years old by the old dating which places it about 100 thousand years ago like several of the others.

 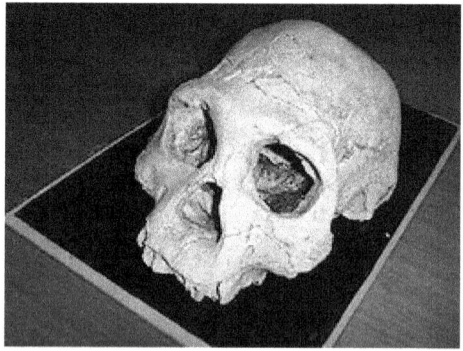

Tools: Some implements and animal bones were found alongside the ancient human remains.

Teeth: Tooth-wear patterns and remains found at the site show H. Georgicus had an omnivorous diet

Fire: There is no evidence of the use of fire.

Brain: At around 600 cm³ brain volume, the cranium was the smallest and most primitive human Hominin skull ever discovered outside of Africa until Floresienesis [hobbit] was found in 2003.

Dimorphism: A strong sexual dimorphism was noted. Males were significantly larger than females. This is considered a primitive trait and is much less obvious in Homo-antecessor, Homo-heidelbergensis and Homo-neanderthalensis type humans. This is one of the reasons to place him in a non-human lane or human-ish race.

Size: Much smaller than Homo Erectus [about 4 feet tall].

Outside Africa: Georgicus was the first species human to settle in the Middle East as the map shows. By the old dating it was 800,000 years before erectus. By new dating it was still 80 thousand years ago.

Skeleton: Four fossil skeletons were found, showing a primitive skull and upper body but with relatively advanced spines and lower limbs, providing greater mobility.

Classification: While Georgicus is sometimes considered a variant of Erectus, the skull showed shows marked differences. He had a smaller brain, but his skull ridge seemed to be much less pronounced.

Ergaster Race

Homo Ergaster is believed to have been the first human by some. As such, he would have been the man described in the Genesis story that appeared during the 6th age after the end of the Cretaceous period and the "out of Africa Story" would start here. As shown below Homo Ergaster, was possibly the beginning of the Negroid Race. Certainly, it has come through many changes, but the out of Africa Adam and Eve were from this group.

Homo Ergaster simply means "working man" so I guess you can say there are a number of Homo Ergaster people living today. This guy is extinct and did not make it past the Pleistocene extinction. He lived in eastern and southern Africa during the early Pleistocene, between 1.8 million and 1.3 million years ago as depicted in nuclear decay. Modern timing has him living about 100 thousand years ago.

There is still disagreement on the subject of the classification, ancestry, but it is now widely accepted to be the direct ancestor of later humans including the more well-known Homo Erectus although Ergaster and Erectus lived during the same time so there are many who do not separate them as different. I will identify both, but they might just be cousins without any real difference in mutated DNA. Without the DNA, we are only guessing. Some consider Ergaster to be simply the African variety of Homo Erectus who has been found at many locations outside Africa including Java and China.

Tools: Ergaster and Erectus both used more diverse and sophisticated stone tools than the Habilis. He refined the "Oldowan choppers" [next left] and developed the first "Acheulean bifacial axes" [next "2nd"]. OK; they weren't much, but sharpening stones took a knack.

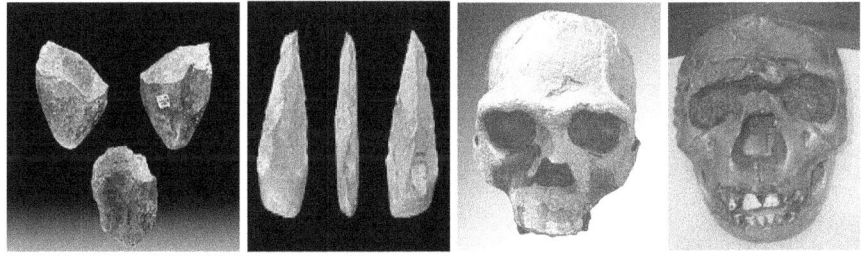

Fire: Ergaster was the first human to harness fire either by containment of natural fire, or as the lighting of artificial fire. This is still a matter of contention.

Linguistic: It was thought for a long time that Ergaster and Erectus were restricted in the physical ability to regulate breathing and produce complex sounds, but now they are believed to have been able to speak.

Anatomy: Erectus is the oldest known early humans . He had relatively elongated legs and shorter arms compared to the size of the torso. These features are considered adaptations to a life lived on the ground, indicating the loss of earlier tree-climbing adaptations, with the ability to walk and <u>possibly run long distances</u>.

Brain: Compared with earlier fossil humans, note the expanded braincase relative to the size of the face.

Teeth and Growth Rate: The most complete fossil individual of this species is known as the 'Turkana Boy', dated <u>around 90 thousand years ago</u> using newer dating methods. Microscopic study of the teeth indicates that he grew up at a growth rate similar to that of a great ape.

Area of Civilization: This guy has only been found in Africa.

Size: Ranges from 5 to 6 feet

Weight: Around 100 pounds

Erectus Race

Homo-Erectus [Erect man] was found first so he has the distinction of being the first human found. Also he is most named the beginning of human race, so we can go with that. Please notice from the chart in the Neanderthal section that Erectus [and Ergaster] were smaller, had a smaller round brains, and were clearly here before the other humans.

The Y chromosome Adam is Haplotype "A". The MtDNA haplotype mutation was "I". This group lived a very long time on the continent and disappeared as the mutated variant we have today took control of the country about 70 thousand years ago. At that time the "B" haplotype and "I1" DNA groups became dominate, Cro-Magnon like beings of Africa.

Well after the last major mutation 5 thousand years ago, this species did infiltrate many parts of the world. By that time, their size, brain capacity and shape, along with the other characteristics of survival had expanded to those of the other competing races of people.

Anatomy: Erectus is the oldest known early humans. He had relatively elongated legs and shorter arms compared to the size of the torso. These features are considered adaptations to a life lived on the ground, indicating the loss of earlier tree-climbing adaptations, with the ability to walk and <u>possibly run long distances</u>.

Brain: Compared with earlier fossil humans, note the expanded braincase relative to the size of the face.

Teeth and Growth-rate: The most complete fossil individual of this species is known as the 'Turkana Boy', dated <u>around 90 thousand years ago</u> using newer dating methods. Microscopic study of the teeth indicates that he grew up at a growth rate similar to that of a great ape.

Society: There is fossil evidence that this species cared for old and weak individuals.

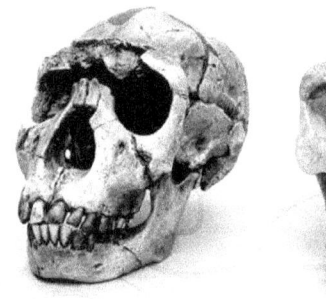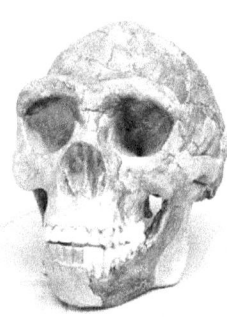

Tools: The appearance of Homo erectus in the fossil record is often associated with the earliest hand-axes, the first major innovation in stone tool technology.

Area of Civilization: Early fossil discoveries from Java and China ('Peking Man') comprise the classic examples of this species. Generally considered to have been the first species to have expanded beyond Africa, Homo Erectus is considered a highly variable species, spread over two continents and possibly the longest lived early human species.

Size: Ranges from 4 feet 9 inches- 6 feet

Weight: Ranges from 88 - 150 pounds

Peking Race

Homo Erectus Pekinensis. Peking Man was a Chinese Erectus with a twist. He lived between 750 and 200 thousand years ago by old timing or about 100 thousand to 80 thousand years ago. A number of skulls were found in a single cave. Inside there was a hearth for cooking and a fireplace of sorts for heating. Not only had he controlled fire, but he set up a home. But that wasn't all. Evidence suggests he made a spear out of wood and stone. It was the first compound tool much earlier than Heidelberg man had made his version. Additionally, Peking guy drilled holes. No one knows why, but he drilled just the same and He liked soft clothing so he softened animal hides.

Peking Man skulls were found in 1929, but sent to the United States to keep them away from the marauding Japanese during World War II. Somehow, we lost them. The sloping forehead and thick brow ridge in front and protruding occipital torus in back are typical Homo Erectus features. Peking Man was a woodworking, fire-using, spear-hafting human-ish Race is still a mystery. His skull is shown below.

Rhodesiensis Race

Homo Rhodesiensis also known as Homo sapiens arcaicus is an extinct human race, described from the fossil remains found in southern Africa, East Africa, and North Africa. **Dating:** These remains were dated between 300,000 and 125,000 years old by nuclear dating or from 90,000 to 75,000 years ago using newer dating methods. They certainly lived during the Pleistocene.

Homo Rhodesiensis is now regarded by some scientists as an offshoot of Homo Heidelbergensis. Originally called "Rhodesian Man", he had characteristics of a new race.

Brain: Cranial capacity of one skull has been estimated at 1,230 cm³.

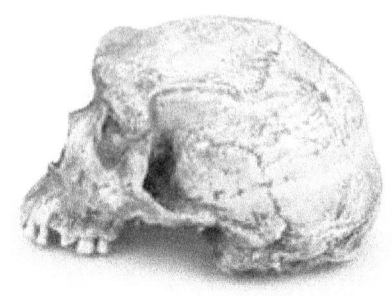

Skeleton: extremely robust individual [especially the skull], and had the comparatively largest brow-ridges of any known human as shown above

Face: It was described as having a broad face similar to Homo Neanderthalensis (large nose and thick protruding brow ridges), and has been interpreted as an "African Neanderthal".

Tool Making: In Africa, there is a distinct difference in the Acheulean tools made before and after 600,000 years ago by this guy as they are thinner, more symmetrical and extensively trimmed.

Teeth: Strangely, one of the skulls had cavities in ten of the upper teeth and is considered one of the oldest known occurrences of cavities. Pitting indicates significant infection before death and implies that the cause of death may have been due to dental disease infection.

Antecessor Race

Called Homo Antecessor these people lived well before Neanderthal sometime around 100 thousand years ago using new dating methods [1.2 million to 800,000 years ago with nuclear decay methods, but generally in the Pleistocene Age like all the rest]. They lived in Europe and they were cannibalistic.

In 1994, 80 fossils of six individuals who may have belonged to the species were found in Atapuerca, Spain. At the site were numerous examples of cuts where the flesh had been flensed from the bones, which indicated that antecessor people ate antecessors. The pictures above show these guys doing just that.

We really don't know if they walked around nude, but there were no department stores and who wants to sell clothing to a cannibal anyway?

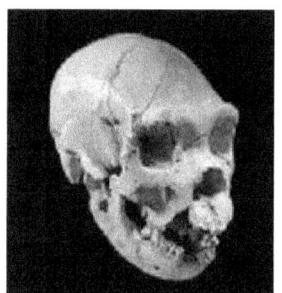 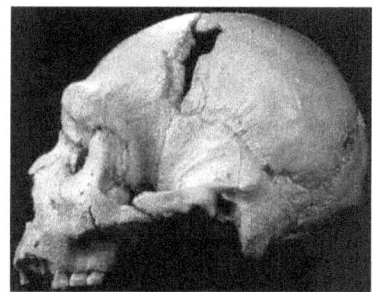

Antecessor Characteristics

Size- These people were about 5½ to 6 feet tall, and males weighed about 200 pounds.

Brain- Their brain sizes were roughly 1,000–1,150 cm³, smaller than the 1,350 cm³ average of modern humans.

Language-Based on tomography techniques it was determined that these people talked with some type of symbolic language.

Face- He had a protruding occipital bun [brow ridge], a low forehead, and a lack of a strong chin.

Where that lived- Remains have been found in Spain, UK, and France.

Tools- At one site in Spain they found approximately 200 stone tools. Stone tools including a stone carved knife. In France were found twenty tools dating back to the time of these people.

Anak Made the Antecessor Race

There is reason to believe the Anak people converted Homo Erectus into this first modification as they tried to increase the brain capacity and make a better "servant". This seems to have been somewhat of an issue as these guys ate each other. One group that the Antecessor possibly tried to eat were called Heidelbergs.

Idaltu Race

Homo Sapien Idaltu is an extinct race of humans of that lived almost 160,000 years ago in Pleistocene Africa using the nuclear decay methods or about 55 thousand years ago with newer dating methods. As a note: "Idaltu" is from the Saho-Afar word meaning "elder" or "first born".

The fossilized remains were found in Ethiopia in 1997. Three well preserved crania are accounted for, the best preserved being from an adult male having a brain capacity of 1,450 cm3. The other crania include another partial adult male and a six-year-old child. Reconstruction is shown below.

Anatomy: Considered the oldest anatomically modern humans, a couple of the skulls are shown below.

 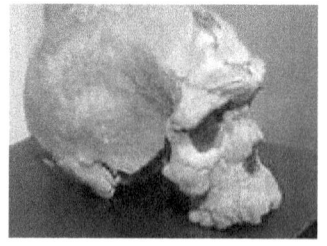

No Burial for these guys

Here is weirdness about the Idaltu people. The child's skull bore marks indicating that, after death, the <u>muscles had been cut from the base of the skull</u>. The rear of the cranial base was broken away and the edges polished, and the entire cranium was worn smooth as if by repeated handling. The second adult skull showed parallel scratches around the perimeter of the skull apparently made by a stone tool repeatedly drawn across the skull's surface in a pattern different from that created during defleshing, as for food. Even the nearly complete adult skull had a few cut marks. It was almost like the skull was worshipped after death. Some believe that the muscles were cut and the skull base broken to remove the brain.

Tools: many were found. More than 640 stone artifacts, and it was estimated that the area of the find contained millions of such artifacts: hand axes, flake tools, cores, flakes and rare blades.

Food: of all things, it seems these guys loved hippo. "

Location: The image below shows where these people lived before the end of the Pleistocene. They did not make it passed the extinction.

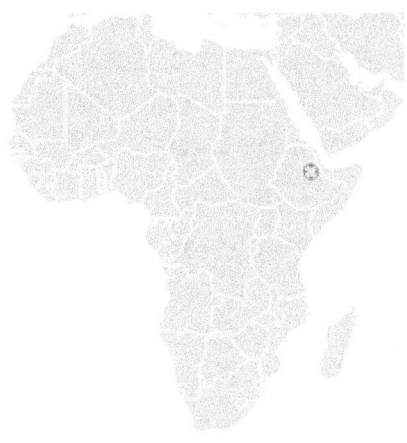

Heidelberg Race

These people were called Homo Heidelbergensis and they lived in Europe well before the end of the Pleistocene and even before the Neanderthal. They have been found with Antecessor remains so there is no telling what conflicts might have arisen. A skull is shown below along with a reconstruction of his face. This guy looked much like a modern man.

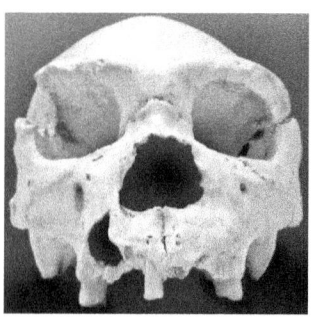

While some have indicated that this race was the first to leave Africa, I don't think there is any reason for the assumptions except that some still want to believe that the Mitochondrial Eve DNA was the first human identified. There are no Heidelberg people who ever lived in Africa.

Some also indicate that these people might have lived as far back as 1.5 million years ago and were the first to talk, build tools, and generally be civilized. As far as we can tell, this race of people died off about 200 thousand years ago well before the time of Adam. Using the newer dating methods, this transforms to about 90 thousand to 70 thousand years ago.

Heidelberg Characteristics

Where – Heidelberg people are mostly found in Northern Spain, but they also have been found in England.

Brain size- It was about the same as modern man and slightly smaller than Neanderthal but larger than their cannibal buddies, the Antecessors.

Body Size – Somewhat larger than Neanderthal, 5 ½' to 6 ½', however, some have indicated that giant Heidelberg people also existed for a time.

What They Ate- We only know what this race ate Ursa Deningeri [an ancient bear] and Mimomys savini [a water vole], but they found a tibia of one of these people that had been gnawed by a large carnivore, suggesting that he had been killed by a lion or wolf or that his unburied corpse had been scavenged after death. We have not found that they had been eaten by Antecessors, but who knows.

Weapons- Hundreds of hand-axes made by these people have been found in England and Spain and 8 wooden throwing spears were found in Germany along with 16 thousand animal bones. From this German site it is believed these guys were very good hunters.

Spanish Colony- It is believed the largest group of these people lived in Spain. So far scientists have found more than 5,500 human bones dated to an age of at least 350,000 years in the Sima de los Huesos site in northern Spain. The one pit contains fossils of perhaps 32 individuals together.

Customs- We know that this race of people practiced complex funeral rites.

Anak Made the Heidelberg-Probably made from the Antecessor, These guys seemed to be much more civilized. It is not hard to believe that these were the first useful servants of the Anak people.

Neanderthal Race

You probably noticed that I didn't discuss Heidelberg Haplotyping. The reason is there is little that is known. While a little more near term, Neanderthal as a race disappeared about 28 thousand years ago and came into existence about 90 thousand years ago using the new dating methods.

 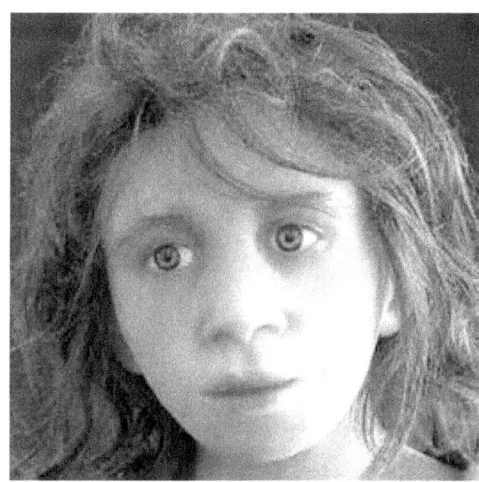

DNA sequencing- With about 15 partial MtDNA sequences to test, scientists have found something interesting as they tested samples from central Asia and all over Europe. There are no similarities in Neanderthal and modern races, but it should be noted that they do not have complete sequences to study. What was determined is that were possibly as many as 4 different sub-races of Neanderthal. The image following shows the main locations where the remains of these people were found.

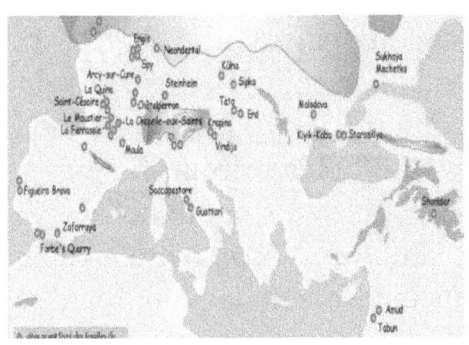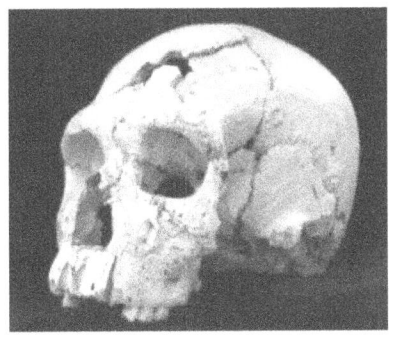

Neanderthal Controversy

There are actually 2 controversial elements to DNA and ancestry here. The first <u>was alien DNA</u> and the second was <u>to whom the Alien DNA probably belonged</u>. An example of the robustness of Neanderthal can be seen in the image previous right. This massively thick skull is of our <u>very smart</u>, very adaptable, <u>larger brained cousin</u>- Neanderthal. Yes; he had a big nose, but he probably liked it that way. He was not a backward race. But he has caused problems when trying to understand races. Usually, the Anthropologists simply ignore him, but let's take a minute to understand his anomalous characteristics so we can use the DNA more effectively.

Neanderthal and the Anak Race

The thing to note about Neanderthal is that his DNA <u>has some alien</u> components. To understand it we must discuss the Annunaki people. Called the Anak by the Jewish ancestors, these special people were here before Cro-Magnon and Homo-Erectus humans and they modified DNA with their own DNA. Anyway, to make a long story short, the Anak may have lost the ability to procreate directly and they had some other issues, but they lived a long, long, long time. During the Tertiary period, they experimented and modified animals and people finally making Neanderthal well before Cro- Magnon came along. Wars were often and bad. Many ancient texts describe them and some physical evidence was even found. One of the articles found was a Neanderthal skull with a bullet hole through the head. An additional skull of an extinct Buffalo was also found with a bullet hole. The entry wounds were small in both cases showing the very high speed and the backs were removed as one would believe from the exit of a "modern" bullet. The small, round, clean, smooth hole without radial cracks was found in the skull of a "Neanderthal" man in the early 1920's, but this man died about 50,000 years ago. The skull is currently at the British Museum and was found more than fifty feet below ground level in Rhodesia. A similar hole has been found in an extinct variety of buffalo and on display in a Moscow Museum. It looks like the Anak were good shots.

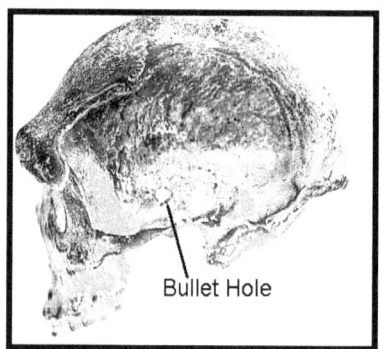
Bullet Hole

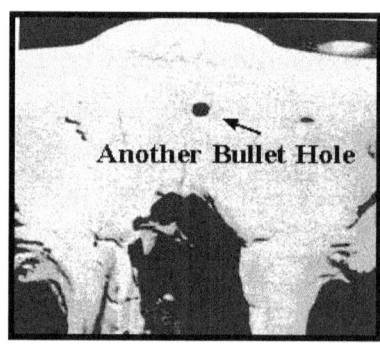
Another Bullet Hole

Anak Controlled the World

Anak or Annunaki [or lords of Annu, or the Arya depending on the society describing their ancestry] were in control of the world in ancient times including the time after the Homo-Erectus had come along. This Homo Erectus was changed genetically. The following graph shows some of the changes that happened overnight as we compare Neanderthal and Heidelberg with Erectus. Denisovan was also added and discussed later.

Neanderthal	Denisovan	Heidelberg	Erectus
Lived until-28,000BC	60,000*	65,000*	65,000*
Found "Alien" genes	Yes	Yes	??
Brain size to 1600cc	1400cc	1400cc	800cc
About 5 ½ feet tall	5 ½ ft.	5 ½ ft.	3 ½ ft.
Elongated Brain**	Similar	Similar	Rounder
Red Hair and freckles	Dark	Light	Black?
Much less hairy	Less Hair	Less Hair	Hairy
Light skin	Dark skin	Light skin	Black?
Lived in Europe	Eurasia	Europe	Africa
Made and used tools	??	Yes	no
Began to bury the dead	??	Yes	no
Made and used jewelry	??	Yes	no
Began to live in villages	??	Yes	no
Began to protect the sick	??	Yes	no
Began to have religion	??	Yes	no

*using newer timing   **Increased motor skills

Anak Convert Erectus Into Neanderthal

The changes certainly were massive mutations, seemingly impossible to have happen. In a twinkling of an eye, so to speak, Erectus had become Neanderthal. To understand this change we need only read a small section in some of the ancient texts to expand on what I already presented earlier.

Enoch 2:18-Three [Anak] came down and copulated with women and had offspring. [The "three" probably indicates three successive attempts at inbreeding with humans or three simultaneous inbreeding attempts.]

Nag Hammadi Creation Text-Now come let us [the Anak] lay hold of her [the human female] and cast our seed into her, so that when she becomes soiled she may not be able to ascend into the light-rather she whom she bares will be under our charge. [The Anak had sex with or modified the Homo Erectus and later they modified Adamic humans.]

Nag Hammadi- [The ruling Anak of the world said] "come let us sow our seed in her [human female]" and they perused her *and she laughed at them for their witlessness and in their clutches she became a tree and left her shadowy reflection- and they defiled it foully.* [The Anak were not always successful at impregnating humans.]

Melchizedek-*Pray for the offspring of the angels, together with seed which flowed forth from the father of all who made the entire universe from nothing there were engendered the gods [ANAK people] and angels, and the men that came out of the seed, all of the natures, those in the heavens and those upon the Earth—now the nature of females was wanting among those that are in the heavens. They were bound with men and women, but* <u>these were not the true Adam</u> *nor the true Eve.* [This verse talks about a difference between angels and people called the ANAK and infers that a union between man and one of Anak was accomplished. It specifically indicates this [Half-breed or Gentile human was not the true Adam and Eve.]

Middle Eastern Changes recorded-The Sumerian, Indian, and Iranian, descriptions were the most compelling. *After having sex with one of the Annunaki people, the man was no longer to be associated with beasts.* <u>His hair was no longer all over his body</u> *and his legs were diminished. He became a hybrid.* [I think we will someday find that Homo-Erectus had hair all over its body while we know Neanderthal had a light complexion and red hair.]

India Creation Myth-*The first humans were covered with **thick hair**. When they mated, they produced people as they are now.* [We can assume they mated with the Annunaki people.]

Sumerian Gilgamesh Story-*Aruru [one of the Annunaki] pinched off some clay and created a [primitive man] Enkidu. His whole **body was covered in hair**. He knew neither people nor country; with cattle, he quenched his thirst, a hunter and brigand. She [Shamhat –one of the Annunaki] must take off her clothes and reveal her attractions. Do for the primitive* *man [Homo-Erectus], as women [Annunaki women] do. She pulled not away, Enkidu was aroused.* [This is the most descriptive and shows that there were female Annunaki that had sex with male humans. They sort of forced themselves to sleep with this lowly being.] ---*Afterward- the gazelles saw Enkidu and scattered, for Enkidu had stripped--- his body was too clean [the hair was all gone]. His legs were diminished; he could not run as before, he had become wiser.—Enkidu, you have become like a god [Annunaki human]. He shall bring up daughters of gods [what we might call Gentiles.* [The Union between the Annunaki and the new human produced viable offspring.]

Mandaeans of Iran Story-*According to their traditions, god first made man. When he was finished, he looked like a man, but moved on all fours, had the face of an ape, and made noises like a sheep. Only later did he put in a soul and teach him and make him erect.* [This ape-man was not Adamic man, but Homo-Erectus.]

African Erectus Mating Story-In Africa, the same story was told. *Hairy men became human after coming in contact with Annunaki.*

Southeast Asian Erectus Mating Story -In this story, we find the same thing. *An extremely hairy human "female" named Bota Ili was cooking food. A non-hairy fisherman named Wata Rian [one of the Annunaki] saw her and got her drunk. While she was asleep, he shaved her entire body. Only then did he find out she was a woman. She learned to wear clothes, they married, and they began a new race.*

Emerald Tablets [Egypt]-*The master said-take them by the arts ye have learned of far across the waters until ye reach the land of the **hairy** barbarians, dwelling in caves of the desert. Follow there the plan.* [The plan was to inbreed with the hairy barbarian Homo-Erectus.]

Ngombe Tribe-See if the Ngombe tribe folklore doesn't sound familiar. "A sky person [Annunaki] saw a hairy man [Homo-Erectus]. She married him and removed his hair [He became Neanderthal]. Then a Garden was made for man to live in." [The offspring of this union was less animal like just like all the other descriptions.]

American Homo- Erectus Story-In the Americas, the same story was told. *"Men were once hairy and monkey like." Something happened to them.* [I think you know what it was. It had to do with Anak and sex.]

Inca Legend-*The age of primitive man [hairy?] was before the Age of Heroes* **[Annunaki and demigods]** [Primitive man turned into heroes probably by a process called breeding.]

Aztec and Mayan History-According to Codex "Laticano-Vatino" the Homo erectus, man was hairy. *During the age of the four winds men turned into monkeys---* [This may be a reference to man with thick hair]

Greek Homo Erectus Change-The Greek Mythology was not as specific, but the Greeks did indicate that the early attempts at gods having offspring with man resulted in monsters. Here is what the Greeks said.

*"First only existed chaos and Gaia which would be the Earth was formed. Gaia gave birth to the sky. The gods Gaia and Uranus had children-12 were Titans, 3 were Cyclopes **and three were monsters with 100 hands. One of the titans took control [Cronos] and created the gods** who took control by freeing the monsters. Zeus had a woman created from clay [Pandora]. She opened the box of knowledge and released evil into the world." [Later Zeus became a lecher with human women.]* [Anak and humans mated. Actually, those called Titans were here before the Anak, but in this book, I have lumped them all together for simplicity.]

Review

The Anak had converted Homo Erectus into the Neanderthal. The Hairy miniature, small brained man had changed completely.

Previously, Neanderthal was described as still being naked and hairy with a shaggy black beard and massive nose as shown to the left above. DNA tests demonstrated that Neanderthals possessed fair skin and reddish hair. They had a high level of "Rufosity": having reddish hair, with red pigments, or natural freckles. We can believe the Anak also had many of these traits, so that part of Neanderthal probably came from them. We will see these things come up again in the Red Nordic races. Evidence of clothing, tools, hunting skills,, burials, handmade jewelry, toys, and other things allow us to know this was a well-developed and social group of people. I spent some time on Neanderthal as he almost seemed to be the highest evolved man before Cro-Magnon came along, but then we found Denisovan. Similar to Neanderthal, this race of people had some anomalies.

4 Sub Races

We tend to think of Neanderthals as one species of cavemen-like creatures, but now scientists say there were actually at least three different subgroups of Neanderthals. Using computer simulations to analyze DNA sequence fragments from 12 Neanderthal fossils, researchers found that the species can be separated into three, or maybe four, distinct genetic groups as those in Western Europe, Southern Europe, Eastern Europe, and the Middle East, have DNA differences.

Art

One sign of civilization is Art. What we have found is that jewelry was made [below right], musical instruments and even the occasional carving on the wall of a cave as shown next left. While it doesn't look like much, these guys were free thinkers and creative.

Anak Made the Neanderthal

Like those before, we can believe the Anak Race was responsible for the "update" in Heidelberg people. With more creativity came more capability. As far as we know, there was no major upheaval that forced massive DNA mutation as we see after the worldwide flood. Somehow a new race of humans came along and with them there was mystery.

Denisovan Race

Let me let you in on a dirty secret. If you want to have a viable community, you must have many opportunities for procreation. Some don't work, sometimes location is a huge deterrent, sometimes the offspring are not viable, etc. etc. That being said, tracing back family lines to a single group from a single parent is not likely. I'm not saying there will not be similarities in how each of the potential offspring producers would mutate, but to think that all would mutate the same is--- OK I'll say it again; UNLIKELY. Somehow the Denisovan Race didn't get the information and messed up some of the Haplotype studies. Denisovan people seem to be a cross between Heidelberg and Neanderthal people; in fact, they are more closely related to the Heidelbergs according to their MtDNA. Recreations are shown below.

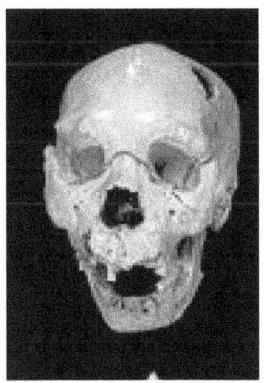

A higher quality Denisovan genome published in 2012 revealed variants of genes in humans that are associated with **dark skin, brown hair and brown eyes** are contained in this race showing he looked substantially different than Neanderthal.

Not Out of Africa

I know you keep hearing a new mutation of people came out of Africa 200 thousand years ago, but it simply didn't happen or the probabilities are extremely low except for the Homo Erectus people. Certainly, there were early people in that country, but there were other people in other locations as I have been presenting. One race of people was found in the Far East and SOMEHOW these guys mated with a European Heidelberg from Spain. The map below shows where a finger bone of Denisovan was found in Russia.

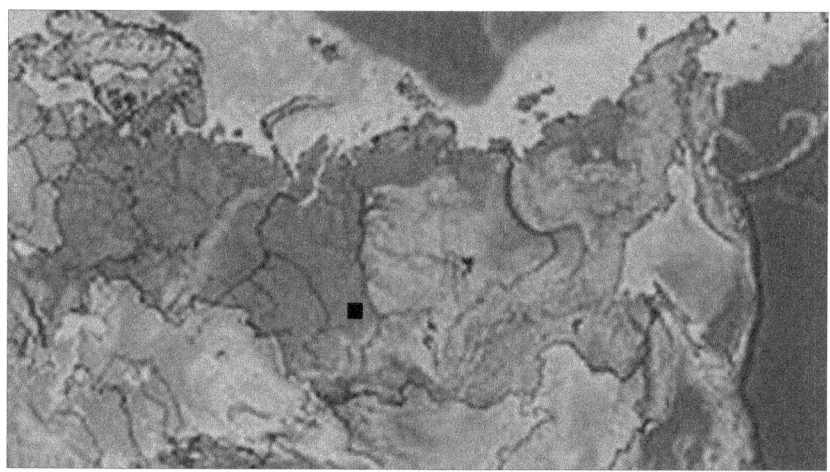

Their DNA is the troubling thing. Let me show you how they try to resolve DNA connections. In the following graphic, you will see Denisovan are genetically similar to Melanesians [Australians and New Guineans] and that is simply not good because Melanesians are directly related to the various Cro-Magnon races. Also, notice that Neanderthal and Heidelberg DNA must be linked oddly to make sense of it as well. No one can figure out how this is possible, but there it is.

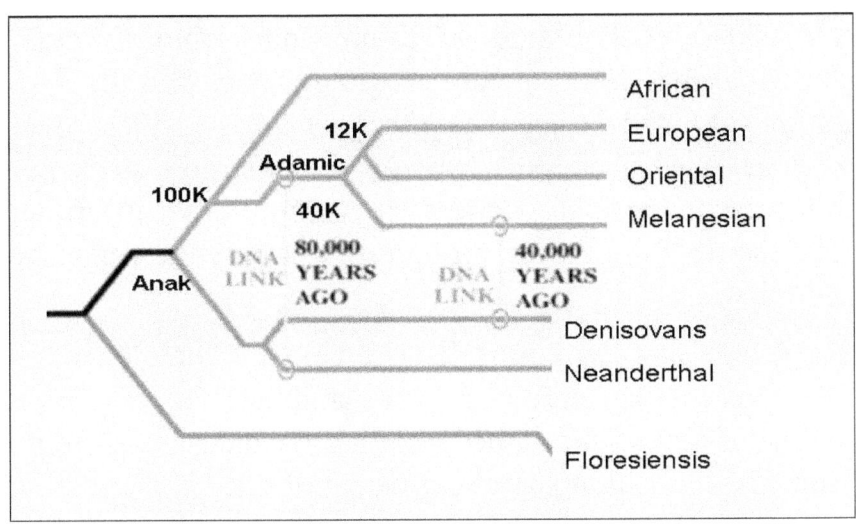

Denisovan Mutation

Denisovan were essentially Neanderthal that were not located with Neanderthal. Completely isolated from the REAL Neanderthal, some of their DNA changed during a mutation period differently than the European version.

Wrong Denisovan Tracking

An international group of scientists have completed a highly detailed analysis of DNA from what is estimated to be a 50-80,000 year old Denisovan finger bone. Common thinking from anthropologists without insight now say about 700,000 years ago [about 90 thousand years ago by new dating], a group of humans left Africa and spread out across Europe and Central Asia. Those walking the thousands of miles to the Far East and those walking to Europe stayed the same somehow [their DNA did not mutate]. Here is where we must say there are no Neanderthals in Africa so as these people left, both groups were simultaneously zapped with cosmic rays or something and both became what scientists call Homo-Sapien-Neanderthalis. During the thousand years of wandering to get to Cambodia and jump over to Australia, there were NO significant changes of either group and the Africans never had a cosmic blast to make any similar people even though there

97

such a HUGE likelihood of the change from Erectus to Neanderthal that 2 completely separate groups had the SAME mutation. Oops! I laughed a little just then, but I'm over it now. This is what happens when a scientist gets so involved with his particular insight and disregards all other information around them. No one even thought to ask how the finger bone got to Siberia.

Besides this guy that must have taken a wrong turn to go North, soon they go tired of walking and had kids. The kids of both groups stayed the same. Both are Neanderthal except for the unusual DNA that somehow was introduced. These people became Neanderthals and Denisovan, and the people they left behind in Africa or the Middle East became Homo-Sapien-Sapien. So how does a Denisovan finger bone illuminate what happened to Neanderthals? <u>One thing that we now know is there are 23 known areas of the genome that modern humans do not share with either Neanderthal or Denisovan.</u>

Flying

Another thing that is almost certain is that flying transports were common in the olden times allowing for widely spaced separation of individual, societies, and trade. Hundreds of document, physical evidence, and artwork describe them, explain how they were used and allow us to understand more about DNA if someone would simply try to explain something in a less comical way.

Cross-breeding

Scientists have now found a crossbred individual "half Neanderthal and half Modern man". When the DNA structure of Europeans is examined, it is found that they now are between 1 and 4 percent Neanderthal. When I say cross breeding, I have to also add in Denisovan.

Oldest DNA

In an underground cave in the Atapuerca Mountains in northern Spain, a pile of bones was found. In the pile was a leg bone with DNA. The bone is apparently 400,000 years old [old dating] or about 50 thousand years old [new dating] and shown below.

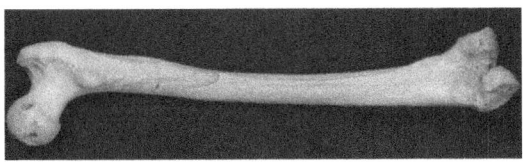

The researchers reconstructed a nearly complete genome of this fossil's mitochondria. The fossils unearthed at the site resembled Neanderthals, so researchers expected this mitochondrial DNA to be Neanderthal but to their dismay, the ancient Mother was Denisovan. Now it seems that these Denisovan stayed with Neanderthal and went northwest to get to Spain. Then the Denisovan group was mutated without the Neanderthal getting exactly the same mutation and most of the Denisovan left for Australia while at least the finger of one of them went north to Siberia.

Please don't get me wrong. I don't believe this story. It is what they are trying to make everyone else believe. From this, let's go to something we have details about. We know quite a bit about the beginning of the "8th Age" and the "creation of Adamic people". Scientists call them Cro-Magnon or Homo-Sapien-Sapien. Don't ask me why Sapien has to be written twice.

Inbreeding

In 2010, it was concluded that the Denisovan population shared a common branch with Neanderthals or Heidelberg from the lineage leading to modern African humans. This suggested that the divergence of the Denisovan mtDNA resulted from the persistence of a lineage purged from the other branches of humanity. A detailed comparison of the Denisovan, Heidelberg, Neanderthal, and human genomes has revealed evidence of a complex web of interbreeding among the lineages. Through this interbreeding, 17% of the Denisovan genome represents DNA from Neanderthal population, while evidence was also found of a contribution to the Denisovan from an **ancient human lineage** yet to be identified. I'm not saying this "alien" DNA was Anak DNA-----Maybe I was!

Artificial Mutation

For the average person about 4% of your DNA is known to be from Neanderthal; whereas, about 5% of the genome of Melanesians DNA came from Denisovan. Tibetans have a DNA haplotype mutation that assists with adaptation to low oxygen levels at high altitude and in 2014 it was found that **Denisovan had the same mutation** showing their close relation. With the massive distances between each of those that seemed to be connected with Denisovan, it is easy to conclude that the Anak were working to modify and make Neanderthal more rigorous. One group of Anak must have tried their genetic art in Indonesia.

Floresienesis Race

As Homo Erectus wandered in the Orient, there was a mutation that occurred about 70 thousand years ago [using new dating system]. This new mutation was called Homo Floresienesis (nicknamed 'Hobbit'), have been found between 50,000 and 17,000 years ago on the Island of Flores, Indonesia.

Size: 3 ½ feet tall,

Weight: 30 kg (66 pounds) - estimate from a female skeleton

Brain: had tiny brains,

Teeth: large teeth for their small size,

Posture: shrugged-forward shoulders,

Chin and Face: no chins, receding foreheads,

Feet and Legs: relatively large feet due to their short legs.

The images following first show Floresienesis skull compared to a modern skull and the second shows that even with most of the animals on the island being miniaturized, they were still much larger than Floresienesis.

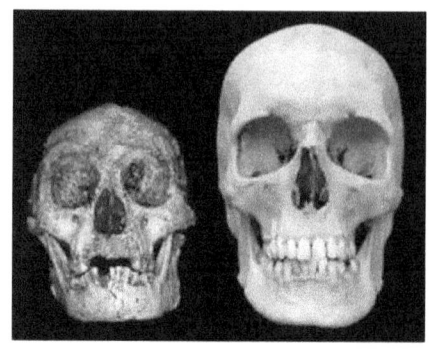

Tool Making: Despite their small body and brain size, they made and used stone tools,

Occupation/food source: hunted small elephants and large rodents, coped with predators such as giant Komodo dragons.

Fire: may have used fire.

Haplotyping: from limited DNA details, Floresienesis are blocked as LB-1 Haplotype. Separating him from Homo Erectus.

Why Were They Small?

It seems everything on the Island got small. Potentially some external mutational element existed here, but it is not clear. Pygmy elephants on Flores, now extinct, showed the same adaptation. The smallest known species of Homo and Stegodon elephant are both found on the island of Flores, Indonesia.

End of the Race

As Floresienesis did not go north towards the Cro Magnon fighters, they survived when the Neanderthal died out.. It is apparent that they did not survive the extinction period at the end of the Pleistocene Age but it probably was more about their tiny size. Unfortunately for other races, the Cro-Magnon came along and may have been responsible for killing off a number of the ancient people. Let's look at Grimaldi.

Grimaldi-man Race

Cro-Magnon's struggles did not end with Neanderthal. It has been shown by mtDNA traces that afterwards, other races entered Europe, specifically, and the Congid [pre-negroid race] tried to take their piece of Europe. As the pre-negroid "Grimaldi-Man" shown below lived, fought, and eventually died off in southern Europe between 40 thousand and 10 thousand years ago. As the Cro Magnon was advancing, it seems this group outlived the Neanderthal. Grimaldi was established as the Cro-Magnon and Neanderthal had mixed with Homo-Erectus to establish this briefly lived race.

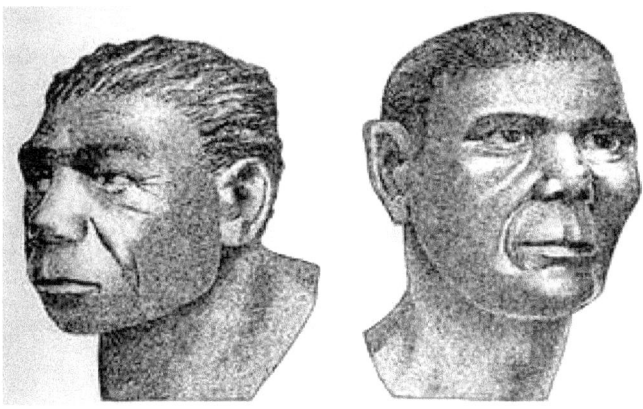

In the early 20th century, two Paleolithic skeletons were found. The skeletons differ markedly from the contemporary Cro-Magnon finds from other parts of Europe, the Grimaldi find was originally classified as a Cro-Magnon offshoot, the features were substantially Negroid. One of the two skeletons belonged to a woman past 50, the other an adolescent boy.

Grimaldi-Man Characteristics

Build: The Grimaldi skeletons were somewhat slender and gracile, even more so than the Cro-Magnon finds from the same cave system.

Size: The Grimaldi people were small. While an adult Cro-Magnon generally stood about 180 cm tall neither of the two skeletons stood over 160 cm [5 ' 2"]. The boy was smallest at a mere 155 cm.

Brain: The skulls of the two had rather tall braincases, unlike the long, low skulls found in Neanderthals and to a lesser extent in Cro-Magnons. The cranial capacity was also quite large for their size [about 1580cc].

Nose: The faces had wide nasal openings and <u>lacked the rectangular orbital and broad face so characteristic of Cro-Magnons</u>. The nasal bones gave a high nasal bridge, like that of Cro-Magnons and modern Europeans and very unlike more tropical groups.

Muscles: Total muscle mass determined showed the two would have been well muscled in life, rather than having the slender build usually seen in tropical people

Family: As shown below, family members were buried together to show their closeness, family unity, and close-knit society.

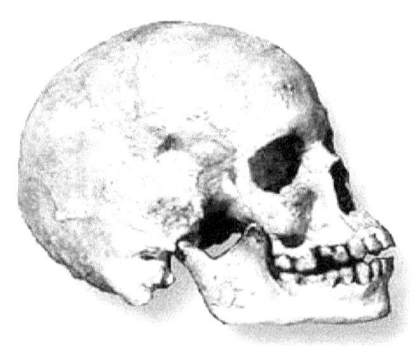

Art: Fine artwork has been found in caves occupied by these people. One such artwork is shown below.

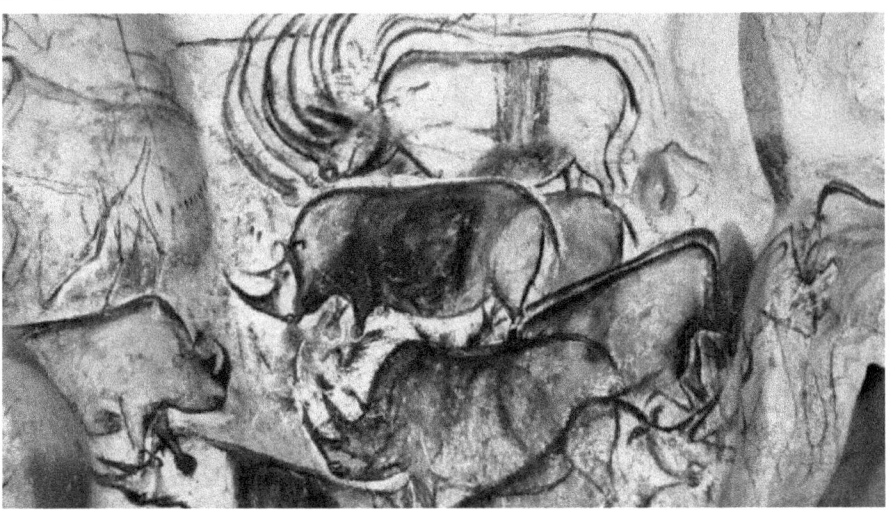

Boskop Race

Homo-Erectus didn't just spread into Italy. Scientist found some strange people that lived during the Pleistocene that had huge brains. Known as the Boskop Man [**Homo-Capensis**] because he was found in Boskop, South Africa, these people must have been "worked" on by the Anak scientists. Reported to have a brain capacity of as much a 30% over a modern human, this guy must have been something. He was certainly pre-negroid, but no one could mistake his brain. Boskop man lived in southern Africa between 30,000 and 10,000 years ago and he evidently did not survive the flood at the end of the Pleistocene. However, similar skull structures have been noted in modern Bushmen or Hottentot people. A comparison of a Boskop Man and modern skull is shown below. The large Boskop is to the left.

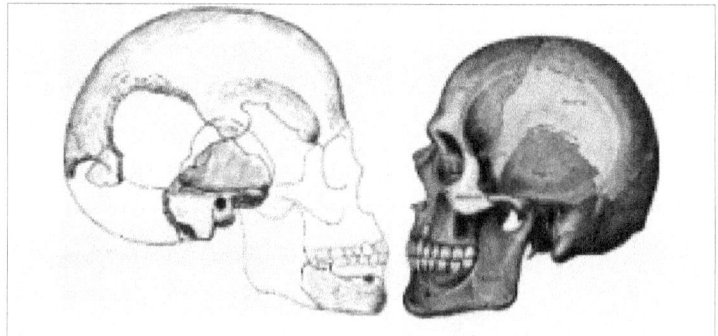

Body Size: The Boskop people were average sized

Skull: Dolicocephalic and 25% larger than modern man. The skull is unusually thick so that the brain size may only have been 1700 to 1800cm.

Brain dimension: Like the Grimaldi brain, it was more of a rounded brain than the longer brain of Cro Magnon.

Nose: Similar to Grimaldi, their faces had wide nasal openings and lacked the rectangular orbital and broad face so characteristic of Cro-Magnons. The nasal bones gave a high nasal bridge, like that of Cro-Magnons and modern Europeans and very unlike more tropical groups. Another image of the skull is shown below

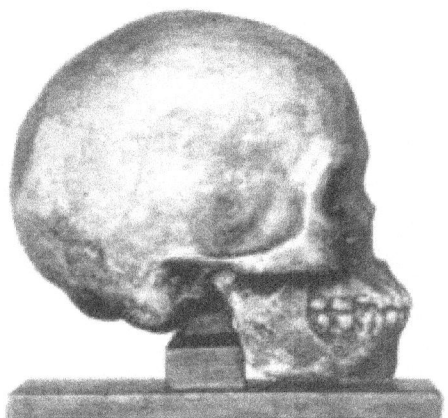 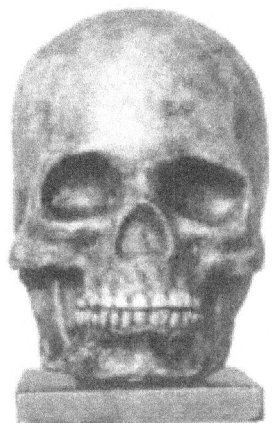

Massive Mutation 40,000 Years Ago

You can call it Mutation, or genetic manipulation, or happenstance, or new God creation in-bred with the help of the Anak people. The thing to note here is that 40 thousand years ago a major change in species occurred and the DNA can track the change and that is not all. There are quite a few reasons we can believe the 40 thousand year marker. Certainly, you were told that Cro-Magnon came about around this time, but what do the other histories say?

Biblical History [4 Ezra 14: 10-12]

For the world hath lost his youth, and the times begin to wax old. For the world is divided into twelve parts, and the ten parts of it are gone already, and half of a tenth part: And there remaineth that which is after the half of the tenth part. [These writings come from about 3.5 thousand years ago. If we assume there is about a thousand years to go before the earth is destroyed and that makes up only 1/8 of the total time the earth has been around, the beginning of man would have been 36 thousand years before the time of Estras or 40 thousand years ago. I know it could also mean that the world itself began 40 thousand years ago, but other data doesn't support that interpretation.]

Babylonian History

King Alalamar [one of the first Babylonian rulers] *ruled for 36,000 years before the flood.* [When added to the 10 thousand years since the flood, this king, [ANAK-Adamic

Hybrids called Gentiles] ruled Babylon about 45 thousand years ago.]

Greek History

Manteo's "History of Egypt"- Gods ruled for 13,777 years followed by 15,150 years of rule by demigods and spirits of the dead before the flood. [When added to the 10 thousand years since the flood, the first man or Gentile ruled Egypt 40 thousand years ago,]

Greek History- According to Diogenes Laertius [3rd century AD], the astronomical records of the Egyptian priests began in **49,219 BC.**

*Greek Historians-*Mochus, Hestieus, Ephorus, Nicolaus, Hesiod, Hecatseus, Hellanicus, & Acusilaus indicated that *the ancients lived thousands of years.* [Details came from Josephus' History.]

Egyptian History

Turin Papyrus- Gods ruled Egypt for 13,420 years followed by 23,200 years of rule by demigods [When added to the 5 thousand years since the papyrus was written, the first man or Anak-Adamic Hybrid ruled about 41 thousand years ago.]

Emerald Tablets- According to this Egyptian text, originally written by Thoth himself, Thoth was sent to Egypt from Atlantis before it sank. He was sent 37 thousand years ago; he became the ruler; and at some point, he created the first 2 pyramids. [Assuming the 30 thousand years started near the time of Adam and Eve, and the reference is about 5 thousand years old, then Adam and Eve would have been here about 43 thousand years ago.]

*Hieronymus-*This Egyptian historian indicated that the ancients **lived for thousands of years**.

Incan History

Popul Vuh- The Inca writing indicates that *the age that followed the "age of primitive man" was the "Age of heroes* [**Titans**] *and demigods". It indicated that 16 sons of the Gods* [**Anak people**] *ruled the land followed by 45 priests* [**Gentiles**]. *After this time, there was destruction by flood and 2 were saved. The survivor sent a falcon to test the land and finally God sent a rainbow.* [Although the time is not indicated to be the 40 thousand years, it is not hard to assume that with life spans of almost a thousand years that this too represents a 40 thousand year era.]

Byzantine History

Syncellus wrote in the 9th century AD, that the chroniclers of the pharaohs had recorded events for 36,525 years.

Roman History

Martianus Capella, from 5th century AD, wrote that Egyptian Sages had secretly studied astronomy for **40 thousand years** before they imparted their knowledge to the world.

Chaldean History

Chaldeans indicated that there was **39 thousand years** between the first dynasty and the flood.

Ancient Text About Races

An ancient text found in the Negev Desert might provide us with some other details about the various races before the worldwide flood time who lived during the time of the Cro Magnon. This ancient text identified 6 types of humans. How they might match with bones that have is not known, but it is more information to give you.

Pleistocene Races

Most of the humans I described so far had died out by the time Cro-Magnon came along 40 thousand years ago, but there still were a number of different races of people including the Anak rulers. Besides color, we may also have an idea of the other physical features of the various people. Portions of scriptures found at the Dead Sea site known only as 4Q186 and 4Q561 show us the first attempt at classifying races of people that survived the flood into at least 7 different groups. The classification was done with respect to percentage of the "light" {or pureness with the Adamic Race or Cro Magnon race as we say today} and percentage from the "dark" {descendants of the Anak's genetic manipulation with Homo Erectus}. Unfortunately, much of the work has been destroyed by time, but a clear idea can be seen from what remains. The "dark" man was a short, terrifying, snaggle-toothed, hairy ape of a man while those who were almost exclusively from the Adam line were medium built, had long thin fingers and toes, had straight teeth and almost no body hair. What hair the Adamics had was curly. <u>A strange thing about the manuscripts is that they were written backwards and in two languages as if the Essenes did not want people to readily understand the documents</u>. Here are the paraphrases and fragmented details-turned into English of course.

11% Adamic Race

4Q186 Fragment 1-*His head and cheeks are fat; eyes are terrifying; teeth are different lengths; hands and fingers are thick; thighs are thick and very hairy; toes are thick and short.*

His spirit has eight parts in the house of darkness and one in the house of light.

Probable 25% Adamic Race

4Q561 Fragment - His body hair is ample; voice is stern and does not strain; hair of his beard is plentiful; neither fat nor thin; short in stature; nails are strong... [The percentage detail of this race and several of the others was not recoverable so it is listed as probable only by characteristic similarity.]

Probable 40% Adamic Race

4Q561 Fragment- His beard is reddish; eyes are clear and circular; hair of his head.

67% Adamic Race

4Q561 Fragment- His Head is wide; chin is thin; body is tall; body hair is full; build is thin but well built; hands and feet are medium length and thin; his eyes are fixed. His spirit has three parts in the house of darkness and six in the house of light. [67% Adamic]

Probable 75% Adamic Race

4Q561 Fragment - His hair is mixed and sparse; eyes are of a medium shade; nose is long and attractive; teeth are straight; beard is relatively thin; limbs in fit condition and medium built; elbows are strong and husky; thighs are of medium bulk; feet are of medium length; shoulders are medium width...

89% Adamic Race

4Q186 Fragment -His eyes are neither dark nor light; beard is light and curly; voice is soft and gentle; teeth are fine and well aligned; size is medium and well built; fingers are thin and long; thighs are hairless; soles of his feet and toes are *even and well aligned. His spirit has eight parts in the house of light and one in the house of darkness.*

To make it easier to identify the differences, I have generated a chart of the characteristics from the fragments and identified how they relate to percentage of Adamic blood. The percentages indicated in brackets were added to show presumed numbers because that was part of the missing information. They were determined by placement in the manuscript and details. The highlighted areas were either not recoverable or simply not mentioned. I have placed possible description generalizations in those positions.

% Adamic	11%	25%	40%	67%	75%	89%
Build	Thick	Med.	Med.	Thin & strong	Med. & fit	Med. & strong
Size	Short	Short	Tall	Tall	Med.	Med.
hands	Thick	Strong Nails	Med.	Thin	Med.	Even/ Thin
Hand length	Short	Short	Med.	Med.	Long	Long
Beard	Hairy	Hairy	Red	Full	Thin	Curly/ Light
Body hair	Full	Ample	Red	Full	Thin	None
Head	Fat	Round	Round	Triangle	Oval	Oval
Teeth	Varied	Spaced	Spaced	Even	Even	Fine
Voice	Stern	Stern	Med.	Med.	Soft	Soft
Eyes	Scary	Round	Clear/ round	Fixed	Med.	Med.
Nose	Fat	Fat	Med.	Med.	Long/ thin	Med.

While this type of characterization is skeptical at best, at least one can recognize that these early people recognized that there were mixed races humans that possessed more of less of the Adamic blood. If Noah's family had been the only breeding group, there would have been no need for reasonableness to the categorization. That brings us to the new way of classifying races. Where no DNA is available, scientist used something similar to this to determine races, but later, DNA allowed for a new understanding of Race.

Cro-Magnon Race [Fm:Nf]

To show DNA mutation Haplotype we can say Cro-Magnon has the male [Y chromosome DNA mutation "F" and the Female [mitochondrial DNA mutation "N"]. Clearly the Cro-Magnon, Homo Sapien Sapien, identified by some anthropological experts as the **_most "evolved" human type ever to have lived on Earth_** came about 40 thousand years ago. Most have no idea how he evolved so quickly. To make it stranger, there were no massive extinctions or astronomical disasters that would have increased mutation when he came along.

He was just there as shown above Y-Haplotype F and MtDNA N. Both brand new distinctions. From this group came the Armenians, White Nordics, and Red Nordics that were the cornerstones of racial distribution around the world. The images of skulls of Cro-Magnon are shown next. Many, due to his unbelievable and sudden beginning also refer to this human as the Adamic man. [Sort of the first Jew, Adam was known to be different that the others in the world at that time. The Homo-Erectus descendants in Africa, Neanderthal in

Europe, Heidelberg people, Antecessor of the UK, Denisovan, and others all would essentially be replaced by this guy or be crossbred.

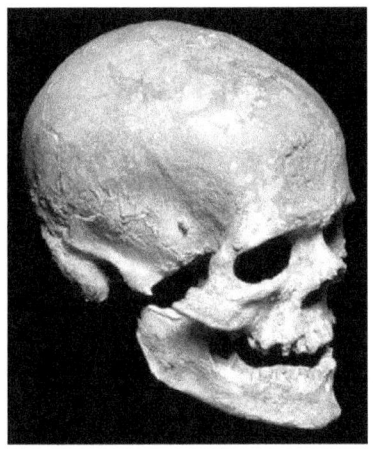 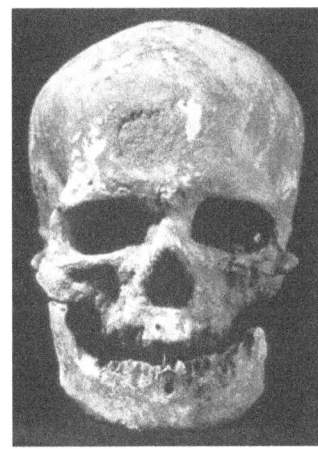

As shown below the northern half of Europe was covered in Ice for much of the Pleistocene as one group of the Cro-Magnon left the Middle Eastern area and went into the Mediterranean area after mutating to [I/Jm:K/Nf] to become the white Nordic Race. Another went towards the Armenian area after mutating to [Gm:Nf] to become the Armenian Race. Another group went farther north after mutating to [Rm: H/Vf] to become the Red Nordic race. Another group went south after mutating to [Jm:Mf] to become the Moslem and Kemetian Race. Still another group went west after mutating to [K/Om:F/Zf] to become the Oriental Race. These and other mutations occurred at the end of the Pleistocene Age.

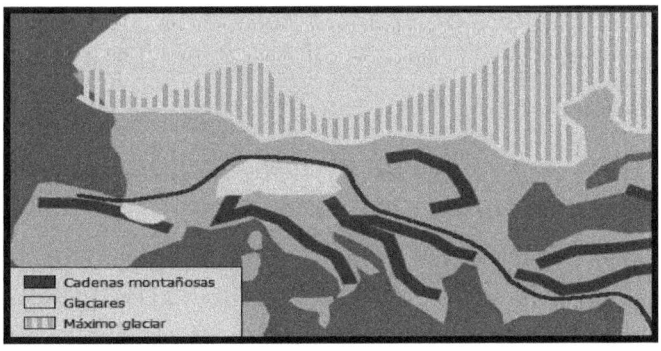

Cro-Magnon Characteristics

Of those identified above, the Armenian race appears to have come about the fastest after the flood. Cro- Magnon had the following features.

Skull: dolichocephalism, a not totally vertical forehead, thin and long skull.

Jaw: squared jaw, sharp chin

Size: tall stature 6 to 8 feet

Face: broad face when compared to modern White Nordics

Muscularity: stronger skeletal consistency, a higher muscular development than White Nordic

Teeth: perfect dental arches

Nose: large nose with a low nasal bridge

Occupation: Hunter gatherer

Tools: The hunter culture showed up with spearheads, arrowheads, richly decorated spear throwers, harpoons, assegais, cave paintings filled with hunting scenes, whistles, and horse's head figures.

Activities: Living a life of violent and constant physical activity outdoors. This developed them as an athletic, graceful, gymnastic human type, and when the climatic conditions got milder, this probably became more apparent. It is in the Cro-Magnon communities of Spain, France and the Balkans, where we have to look for the origin of Greek athletic traditions like races, javelin throwing, hunting, fighting and archery.

War: We know Cro-Magnon had territorial conflicts with reproductive communities of other races. The first race they came into conflict with was the Neanderthal, which had been around in the European continent for 50,000 years and had evolved from earlier populations, such as Homo Heidelbergensis. The sudden disappearance of the Neanderthal has been blamed on climate change and interglacial period adaptation but truly there were many wars and finally the Cro-Magnon triumphed.

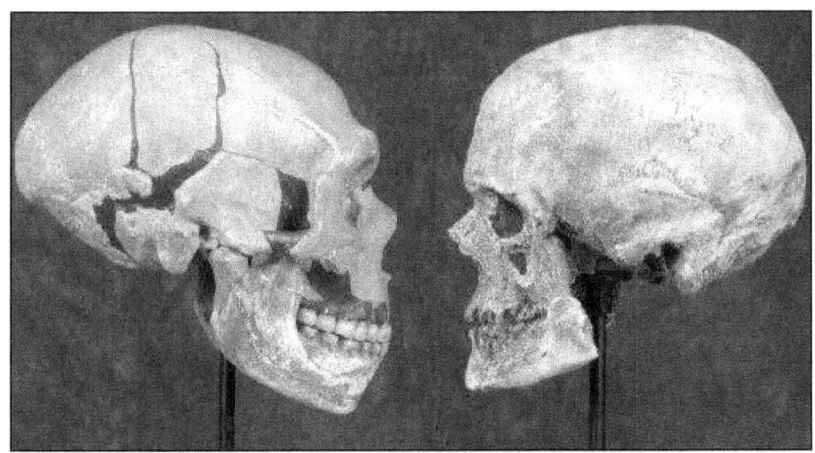

The Neanderthal (left) had been in Europe for over 50,000 years. He had overcome both glacial and interglacial periods successfully and he occupied a territory spanning from Portugal to Central Asia. Despite his fabulous environmental adaptation, when Cro-Magnon (right) showed up, it took Neanderthal only a short time to become extinct.

Ammah Creates Animals

The Anak were not the only ones making animals and new "Men" during the Pleistocene Age. The Cro Magnon people were doing the same sorts of things. I'm sure you have heard about Adam's second wife Eve and possibly even about his first wife Lilith, but this is about 2 of Eve's many children. Timnor and Ammah were their names and they were stinkers and it seems that some of that knowledge fruit tasted by Eve must have trickled down on both of them. Adam and Eve tried to teach their children after they were kicked out of the Garden of Eden, but knowledge has a way of corrupting. While some have placed Adam and Eve's creation about 7000 years ago, new details about the Cro-Magnon "instantaneous" creation and many supporting details place Adam at around 38000 BC. These 2 would have lived around 35000 years ago or so.

Eve and Her 60 Children

According to the ancient Jewish work, "Generations of Adam", the 2 of them had a number of children---60 to be exact. You know some of them, but let me introduce a few others that help us expand this whole strange theme. Eve's first born, Cain, went away with his twin sister, Lebuda, after killing his younger brother or half-brother depending on the text that you read. A couple of Eve's other children were experimenters. According to "Generations of Adam" and other ancient books, they created flying machines, Great War machines, genetically modified animals, and some pretty unusual things. We don't know too much about the other 45 or so children of Eve, but we can imagine that her hands were full. Just imagine being 4 thousand years old and having a 50th child. It probably was old hat by that time and she probably didn't feel a thing but we find that Ammah learned genetics from the Anak scientists.

[6:1-5] Among our little ones was Timnor and Ammah. Timnor understood physical law and created mighty machines. Ammah understood the secrets of creation. She manipulated the very fountain of life until she had created new forms of beings dedicated to the destruction of mankind

[8:4] Timnor and Ammah practiced every abomination. Tranter learned the ways of his mother Ammah and he did manipulate the very nature of man and beast to create new forms which God had not ordained

Notice that Timnor was responsible for building many machines. I'm not talking about a bow and arrow I'm talking about mighty machines that challenged physical law. Ammah had an even greater gift of genetic manipulation. She made all types of animals before the great flood. Many of the animals designed by Ammah were the "Abominable" or unclean animals that were specifically identified as a problem as God hated the idea of people making animals without regard for what they were actually doing. Ammah evidently didn't care. I'm sure Eve tried to keep her from doing these things, but God finally punished the whole world and Noah had to get in a boat for a year while almost everything outside died. Here is some more detail.

"Generations of Adam" 6:2-6

When Timnor was eighteen years of age, he took Ammah to wife, she being fifteen years. Nevertheless, although the Lord Jehovah had blessed these two greatly with wisdom and understanding concerning the secrets of His creation, they followed not after the ways of Jehovah their Eloheim, for they took glory unto themselves, acknowledging not the Source of their power in the heavens. In the one hundredth year, Timnor and Ammah led a group of our children who, like themselves, worshipped the workmanship of their own hands and the power of their own minds, out of the Land of Cainan unto a valley northward which they denominated the Land of Haner, for, said they, here shall we throw off the ways of our Fathers and follow after our own wills. The people of Haner prospered in the land, building mighty cities in which they dedicated themselves to the fulfillment of every physical desire, but their souls were empty because they did not know the Lord Jehovah nor did they call upon His holy name. Yea, Timnor built great engines with which to deface his Mother Earth, --. With other machines of his contriving, his people flew through the air like birds and explored the depths of the lakes and rivers. He created also great instruments of destruction with which his people attacked the people of Cain, --the God-given abilities of men were turned into instruments of death and destruction. Ammah was not one whit behind her husband in creating wickedness, for she manipulated the very fountain of life, until she had created new forms of beings dedicated to evil and the destruction of mankind. From this time, sickness and disease began to spread among the sons of men, bringing sorrow and death upon them. Our hearts were filled with sorrow as we watched the doings of Timnor and Ammah and their brethren, for they were led ever deeper into the depravity of a fallen world.

Ammah Wasn't the Only Geneticist

The Zoroastrian "ZAND-AKASIH" says it this way – *Satan [One of the Anak people] miscreated creatures and they became useless. God saw the defiled and bad creatures, they became abominations to Him. Satan's downfall was the unrighteous creation of the creatures and ignorance*—[By this we can determine that designing animals had been going on well before the creation of Adam and his descendants. The Sumerians indicated that all types of animals were created specifically for the war between the gods.

There must have been all type of DNA baselines in storage or all around.]

What Types of Animals?

Guess what? Scientists are now finding all types of dinosaurs that are not fossilized. Their bone marrow and soft tissues are still ---tissue. This is a problem for many as some believed all dinosaurs were killed at the end of the Cretaceous Period when a meteor hit the Earth. Certainly, they now know that the timing reference used for all extremely ancient dating is flawed. As I mentioned before few years ago, they found out that nuclear decay changes. That being said, it does not account for nonfossilized dinosaurs which are being found around the world today that would have died no more than a 30 thousand years ago. This nuclear decay strangeness does not affect other dating methods so most dating out to 30 or 40 thousand years ago can be generally trusted so these NEW dinosaurs were genetically re-created during the time of Ammah's experimentations.

Clean Animals

She and those who followed her work did not stop there. They began mis-creating everything. The animals that God created the Bible called "Clean animals" while those that had been "modified" were called unclean or abominable" animals. I could give you a list, but it would be too long. Certainly, the Dragon that Daniel destroyed in the last chapter of his book in the Bible was one of the "unclean" ones.

Timnor's Flying Machine

If you ever wondered how the Kangaroo got to Australia in the first place and was saved from the worldwide flood, we can possibly thank Timnor's flying machines. Certainly, there was world trade before and after the flood of Noah using these types of flying machines called Merkaba [by the Sumerians], Vimanas [by the pre-Hindu], Valiaxi [by the Far Easterners], Flying Canoes [by some of the American tribes], and Fiery Chariots [by the early Jews].

Enoch 74:15- *I saw likewise the <u>chariots of heaven, running in the world above to the gates in which the stars turn</u>, which never set. <u>One of these is greater than all which goes around the world</u>.* [This ship going around the world sounds similar to what we do today without space ships, doesn't it?]

Anyway; Timnor's creation possibly sped up the re-colonization of the world. Here are just a few of the hundreds of depictions.

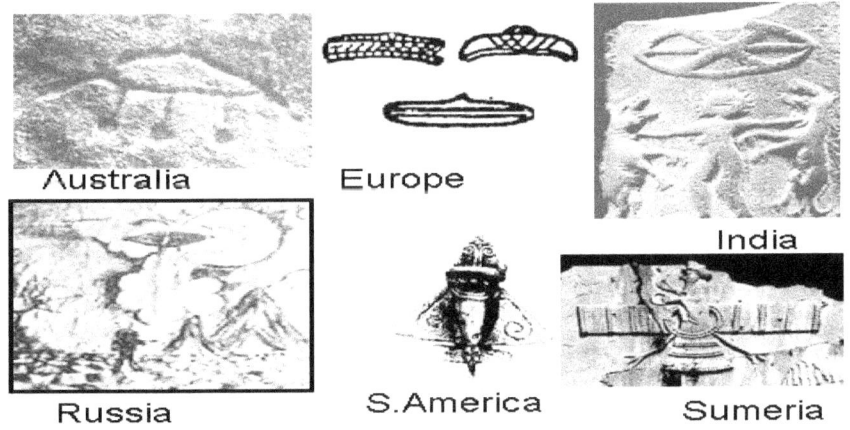

Australia   Europe

India

Russia   S.America   Sumeria

Leboa Daughter of Tamar

After a hundred years or so, Timnor and Ammah got married and had a bunch of kids. Each of them took up one of the more mystical areas of study and they had children. Soon there were thousands and they went to war with the thousands of people from the land of Nod where Cain had become King. The following again comes from "Generations of Adam". One of the descendants of Ammah was named Onesima. Onesima's great granddaughter would turn the ties of war. Her name was Leboa. She was just like her great granddad Timnor. Timnor had developed war machine, submarines, flying ships and things that defiled the very earth. Leboa wanted to help the warring people of this time even more. This section of the book is around the end of the Pleistocene and Wars were everywhere.

2:4- In the process of time, six sons were born unto Timnor and Ammah.

7:3-6- The people of Timnor and Cain began to come upon our children and stole herds and produce, killing any who sought to prevent them. We [Adam and some of his children who stayed true to God's ways] left. Many stayed, determined to fight for their homes.

The wars simply got out of hand. Let's see some more.

8:13-14- A bright light shown from heaven illuminating the whole palace of king Canaan, and a mighty noise from heaven shook the air. Whence the palace stood there was only dust. A war broke out among the people. Armies devastated the land.

The people suffered great desolation. There was destruction everywhere.

__10:2-__The king of Canaan directed his people in erecting great barriers of power around the city of Haner, so that none could pass into the city-neither could any missile penetrate the forces which surrounded the city.

They Didn't Know About Leboa

__10& 11-__ He directed his people in erecting great barriers of power around the city of Haner, so that none could pass into the city without the consent of the inhabitants thereof. Neither could any missile penetrate the forces which surrounded the city. Thus were the people of Palai amazed, when they came upon the city of Haner, for all their devices could do no harm to the inhabitants of the city who remained secure behind their fortifications. Nevertheless, the <u>people of Palai were expert in many devices, and there was one among them, even Leboa the daughter of Tamar, the daughter of Rachel of the House of Onesima, who devised a sword of light which penetrated the wall of defense around the city of Haner and began to drain the power from the wall.</u>

The outcome this fighting, according to the book of Jasher was that about 1/3 of the people died even before the great flood that ended the Pleistocene. Possibly they are talking about the Neanderthal as some of the combatants, or the Grimaldi, the Denisovan, or even the Heidelberg. Bloodshed was everywhere as many people essentially became extinct before the extinction. Luckily, Adam had died by this time living only 5500 years according to the book of "Adam and Eve II". Possibly the wars were still going on when meteors began falling and fires were everywhere.

Massive Mutations 11,000 Years Ago

So we have the Anak, Homo-Erectus, Heidelberg, Neanderthal, Cro-Magnon and all the rest running around the world. While it doesn't seem like it, during the Pleistocene there were not many mutations. People simply stayed people. Sure, there were black ones and white ones and red ones, but there was also a lot of similarity. Overnight as the Pleistocene ended, massive mutation disrupted and huge changes occurred to the DNA.

A study dating the age of more than 1 million single-letter variations in the human DNA code reveals that most of these mutations are of recent origin. **Over 86 percent of the harmful single nucleotide mutations arose between 5 and 12 thousand years ag**o. Oddly, since then, there have been very few of this type mutation. Overall, the researchers now believe that about **81 percent of the single-nucleotide** variants in the European sampled and **58 percent in the African DNA sampled** arose **in the past 5,000 years**. In the African samples a large number of the single nucleotide mutations appeared over 50 to 100 thousand years ago [by standard dating methods] which would be from 40 to 65 thousand years by new timing.

11 thousand Years Ago

It seems odd that someone doesn't come along a state the obvious. That there were huge periods between 12 thousand and 5 thousand years ago where something bad was happening to the environment to allow for massive mutations. While not the reason for this book, here are some of the things we believe happened. Ten to 12 thousand years ago the moon of the planet Venus [Planet Rahab according to Biblical testimony] exploded sending down massive numbers of meteors that caught the earth on fire in many places and shifted the earth's axis to instantly freeze Mammoths while they still had flowers in their mouths. The timing of this last shift can easily be seen by looking at the sharp turn [30 degrees] of the Hawaiian Island hotspot. [See below]

The entire eastern coastline of the USA was peppered with hundreds of thousands of craters from the incident. Today they are called the Carolina Bays but the upheaval allows dangerous radiation to modify DNA. A couple images below show how filled with craters, the eastern Coastline of the United States is still today as over 500 thousand pieces of the moon struck the earth setting fires and momentarily shifting the earth. This set up a chain reaction which eventually shifted the earth to a new rotation we have today. It almost caused complete extinction of the planet. Those who survived were changed as the cosmic rays no longer were held off by our atmosphere.

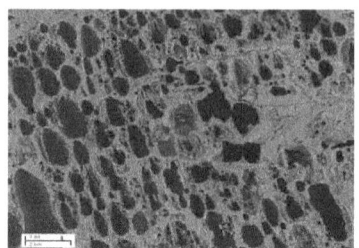

10 Thousand Years Ago

Climate date from Ice core samples tell us about the misery. The earth shifted and the axis was not stable so the earth went back close to its original rotation over a period of a few months. The chart below shows the distinct and abrupt climate change for a very brief time as the worldwide flood covered the Earth then receded.

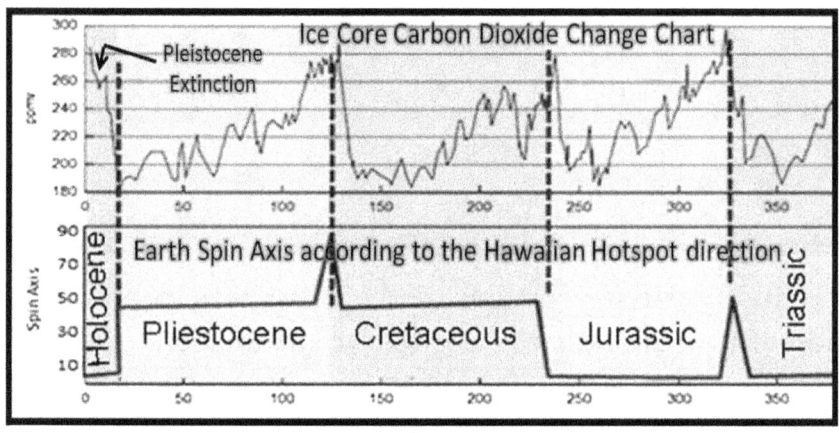

Massive tidal waves and rain covered just about everything. The disruption not only killed many, but also DNA mutations were certainly introduced and people were deposited at different locations as the water subsided. Even the Kangaroo found refuge without going across Asia from the Middle East as some would say. When everything settled down. Eighty thousand worldwide flood stories tell us about the survivors around the world. Mutation after mutation started changing who we were. It would not be the last incident, but it would be another 5 thousand years before the next great massive change in DNA would occur.

The study of DNA mutation began to take hold a decade ago to see who married who and whether someone killed another. Soon maps of societies and when migrations were being established by simply looking at specific mutations and how many mutations occurred. The main mutations were called out by a capitol letter and numbers and other letters were added to less dynamic mutations to further identify people the ART of Haplotyping had been established.

Haplotyping

As I have been using this haplotyping to describe certain elements of the human race, I thought I had better give you more information. Ancient texts confirm a large number of races before the end of the Pleistocene event 10 thousand years ago, but so does our DNA. Let's begin looking at DNA more closely. If you are a male, you will have two Haplotypes, meaning Mitochondrial DNA (female line) Haplotype and the Y, sex chromosome, DNA (male line) Haplotype. Haplotyping is the study of mutations in the DNA. Interestingly you can tell the sequence of the mutation by where it is and what was before and after a mutation. By looking at similar mutations, we can determine ancestry and even some detail concerning when particular races were formed. Generally, some pairs of Haplotypes are from various races as indicated below. The Haplotypes are coded. The first letter is the oldest and most important mutation. This is followed by a number, then a letter etc. to look for subgroups. The listing following shows the Y-DNA, then MtDNA, and finally the general grouping of people associated with these special DNA mutations.

Negroid Races

- *Am:(L0)f --Southern Africa*
- *Bm:(L1/L4)f --Middle Africa*
- *Em:(L2/L3)f --Africa wide*

Mongoloid Races

- *(D/O/N)m:(C/Z/D/G/A/B/F)f --East Asia, Siberia*
- *(K/M)m:(B/P/N/Q)f -- Oceania*

Nordic Races

- *(R/I/T/J/E)m:(R0/H/V/J/T/U/K)f--Mediterranean Areas*
- *(Q/C3)m:(A/X/C/D)f--Easternmost Siberia*

By Mitochondrial DNA

Here is another grouping that may also be useful of mtDNA mutation groupings. These people can look like cousins but have different MtDNA mutations as indicated.

Ameridin Races

- *Native Alaskan, Inuit : A/D*
- *North America: A/B/C/D/X*
- *Mexico: A/B*
- *Latin America: A/B/C/D*

Nordic Races

- *Western Europe: H/V/I/J/U/K/T/W/X*

Jewish and Armenian Races

- *Middle East, North Africa: H/V/I/J/U/K/T/W/X*

Negroid Races [L]

- *East Africa: L3/L4/M*
- *West Africa: L1/L2*
- *South Africa: L0/L1/L2*

Mongoloid Races

- *South Asia: N/M/U/B/F*
- *Melanesia: P/Q/B*
- *Australia: N/P*
- *Russia: Z*
- *Eastern Russia: L/Y/G/A/C/D*
- *East Asia: N/M/B*

Anthropologic Grouping

To simply say European is a problem so Europeans were subdivided into 6 major groups using face and body features as follows: [Dolichocephalic means thin head, and Brachycephalic is round head].

- **Nordic:** *high stature, rosy skin, athletic build, straight nose, well-developed chin, Dolichocephalic, fair hair and light eyes.*

- **Dalic**: *high stature, robust and heavily built, rosy skin, blond hair, light eyes (blue, grey or green), Dolichocephalic or Brachycephalic cranium, big mouth and thin lips.*

- **Dinaric**: *high-medium stature, brown skin, slim build, aquiline nose, Brachycephalic, dark hair and eyes.*

- **Alpine:** *medium stature, fair skin, heavily built, Brachycephalic, brown hair, brown or light eyes.*

- **Baltic**: *medium to low stature, fair skin, strong build, Brachycephalic, light hair and eyes.*

- **Mediterranean:** *low stature, brown skin, physical constitution varying usually slender, straight nose, regular features, Dolichocephalic, dark hair and eyes.*

Haplotype Generalization

Now researchers know these classifications are pretty much all wrong as more and more DNA sequencing has given a new picture of races. For Europe we find only 3 main groups.

- **Western Hunter-Gatherers**. *The indigenous population of Europe- Mostly White Nordic race.*

- **Ancient North Eurasians.** *Mesolithic and Neolithic invaders- Mostly Red Nordic.*

- **Early European Farmers.** *Introduced agriculture in Europe- Mostly Armenian*

We will look at these special races later. More and more testing, sampling comparing and sequencing developed this new way of tracing races. The larger the groups were being tested; the more finite the understanding and picture of how races "evolved" by something called mutation. Also, more questions surfaced. One was the odd timing of mutations. Most mutations noted seemed to occur either 11 thousand years ago or 5 thousand years ago. We'll get into the 5 thousand year old mutation event later as well.

If you are confused enough, let's continue. I think seeing some maps will help.

Y-DNA Haplotypes

As I said, by testing large groups, one can map out where individual groups came from, and how the lineage took control of various places around the globe. The following map shows a generalization of this type of Haplotype flow-map. Following the map are general descriptions of the various mutation grouping and a time-period for each event. The relative timing of the events is similar to known tracking, but I have compresses the timing so we can be closer to the ballpark. While we cannot get an exact time, we can determine what mutation comes first so we can adapt the sequence to known events. Hopefully, from my previous discussions it is known that while it is the best thing we have, Haplotyping is not an exact science. There have been so many intermarriages and so many thousands of years, who came first, the Hamite or the Gaelic, can make it difficult to track entities and attempt the generalization of how a person got to be who he is today. Also, note that I confined the African mutations to Africa. While there certainly were extrusions, especially by the "E" grouping, these would not happen for a while.

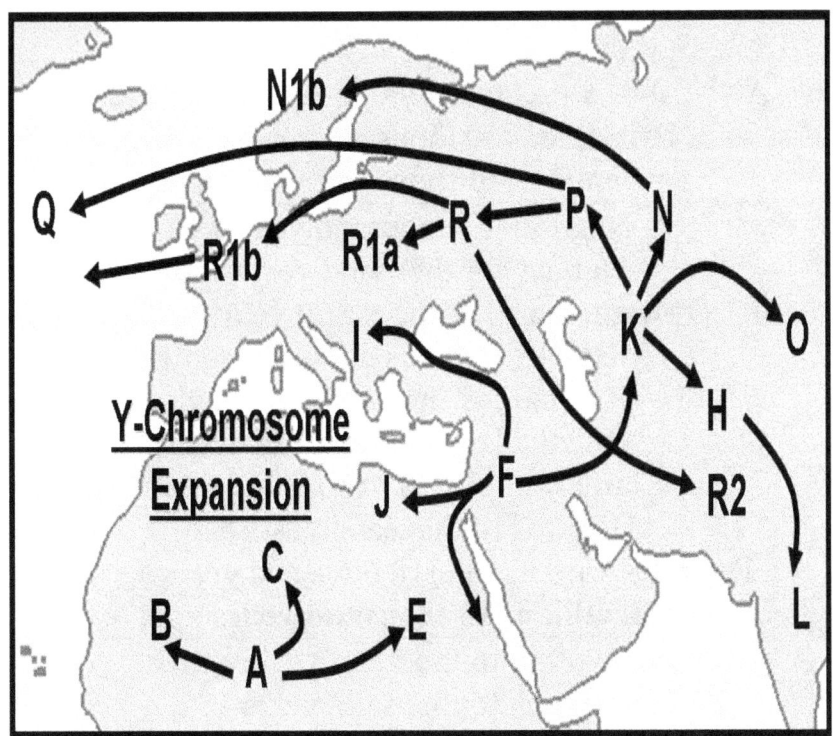

One can define major mutation or combination points by a letter and number identification. The first letter is the most significant ancestral point. It denotes MAJOR mutation points so we can time them pretty well. The number following indicates a single event mutation/modification and a second letter identifies an additional subset or combination of groups coming together. Please note the trail of "major mutation points" is characterized by general characteristics of the groups living in those locations and how they SEEM to flow. For the example above, the string "F to K to P to R" shows a particular timeline. This is believed to be the chain of ancestry for Europeans. Of particular importance is the Haplotyping known as R1b which is the base grouping of Northern Europeans and Eastern North Americans. Sometimes names are given to these groups to make identification easier. The dates are approximations for reference and I have added the names.

135

A= "Y-DNA Erectus" [100 thousand years ago]
B= Sapien [50 thousand years ago]
F= Semitic [40 thousand years ago]

C= Negroid [11 thousand years ago]
E= Nubian [11 thousand years ago]
G=Armenian [11 thousand years ago]
I = Greek [11 thousand years ago]
N= Russian [11 thousand years ago]
O= Oriental [11 thousand years ago]
K=PreAsian [Japheth] [10 thousand years ago]
J=Hamite [10 thousand years ago]
P=Proto-Amerindian [10 thousand years ago]
R =Scythian [10 thousand years ago]

R2= Aryan [6 thousand years ago]
H= Afghan [6 thousand years ago]
L=Dravidian [5 thousand years ago]
R1a=Slavic [5 thousand years ago]
R1b=Gaelic [5 thousand years ago]
N1b= Scandinavian [5 thousand years ago]

Don't worry about all this stuff right now it will make sense as we go along. Also, it should be noted and today some suggest that only the A Genome comes from Africa as the "B" type possibly started in a different country and descendants went "INTO" Africa. The mutation points before 10 thousand years ago would be those that were established before the worldwide flood. The 6 thousand year boundary would be from the beginning of the Babel War and the 5 thousand year mutation boundary was around its ending. To give you a little better perspective, here is a similar chart that shows the major offshoots and when each apparently split off. It also fills in some of the missing lines in the preceding Haplotype map I presented. Please don't be upset about the last 2 human type

Lizard-men and Ape-men. That story is coming up later as those people did not survive long.

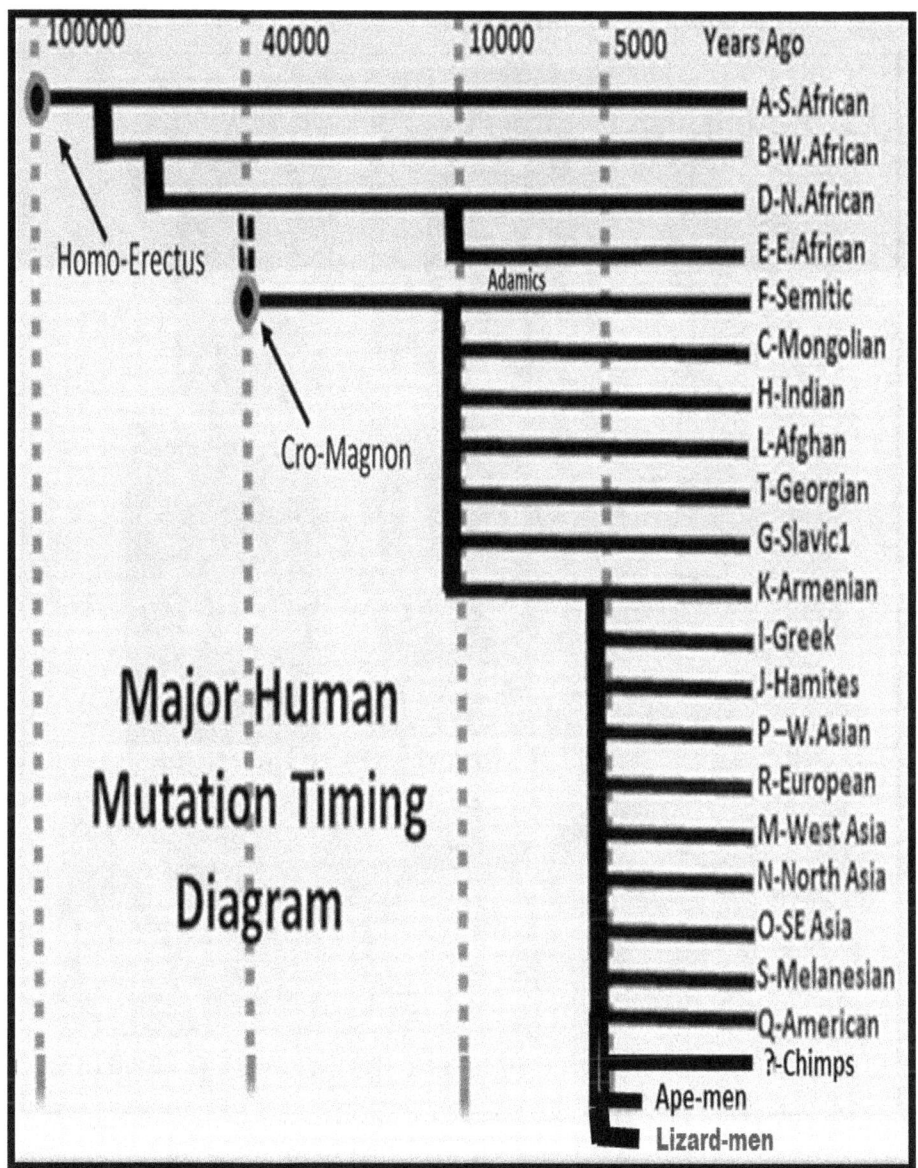

These charts allows us to witness the modifications of the DNA of the Humans that were living so long ago and another DNA sequencing gives us even more data.

Semite Separation

Associated Races are depicted in the following collage. Lower row mutated 5 thousand years ago. The upper mutations were from the Pleistocene Event.

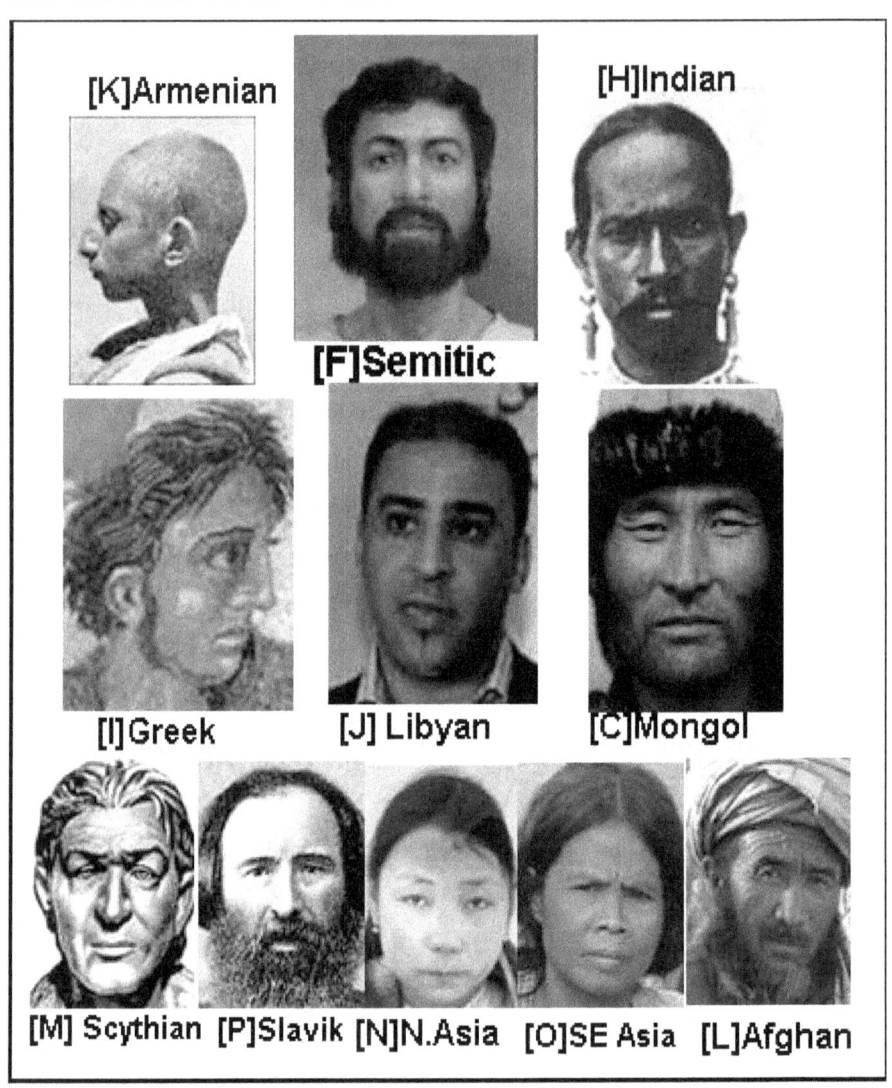

[K]Armenian
[F]Semitic
[H]Indian
[I]Greek
[J] Libyan
[C]Mongol
[M] Scythian [P]Slavik [N]N.Asia [O]SE Asia [L]Afghan

African Homogenizing

Those in Africa, except for a few, remained in Africa, While there were mutations, interbreeding make it difficult to determine various groups. A general accounting is shown in the following collage.

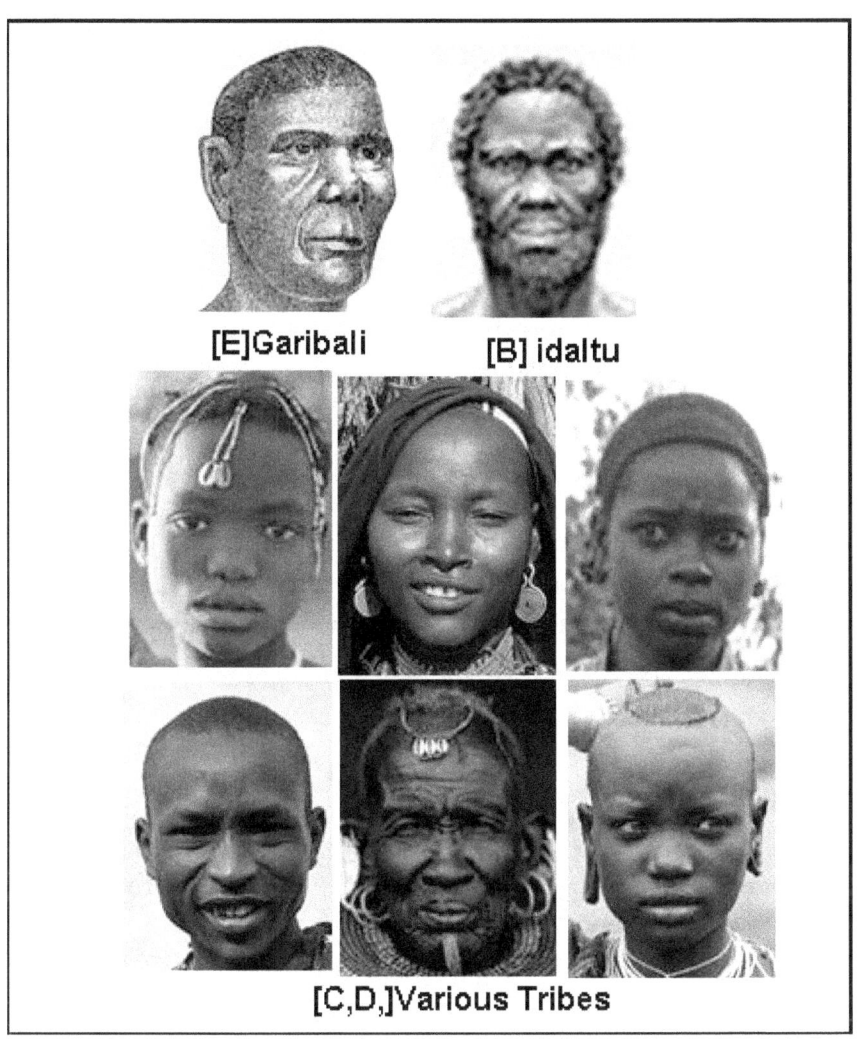

[E]Garibali [B] idaltu

[C,D,]Various Tribes

As before, the upper row is from the Pleistocene, while various tribes had a number of similarities. Various tribe images are shown. Besides tracking by Y-chromosome DNA, the Mitochondria things give us another perspective.

Mitochondrial Haplotypes

While you would think tracking mitochondrial DNA would show almost the same expansion and mutation, this is not the case. One reason might be that the mitochondria are more protected so there are more "straight links". Along the left of the map shows the major MtDNA of the Americas. We will have to investigate how the N and M Haplotypings all of a sudden show up in America without secondary changes.

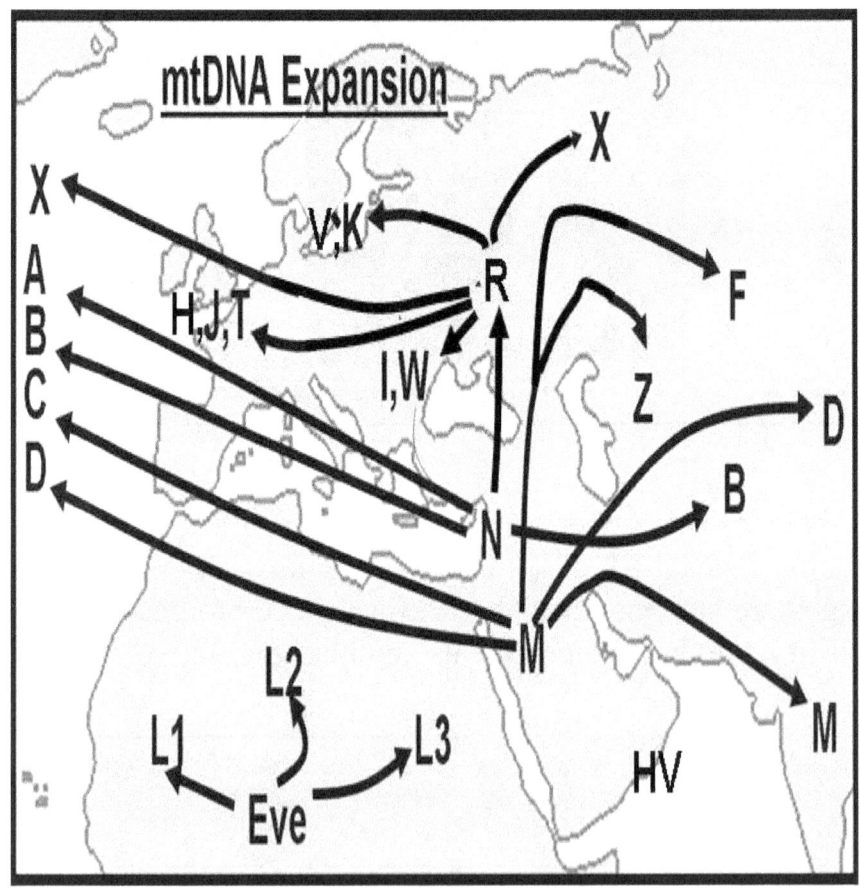

L=Eve=Homo Erectus [100 thousand years ago]
L1= Sapien [50 thousand years ago]
N= Adamic [40 thousand years ago]
M=Arabic [30 thousand years ago]
L2= Negroid [20 thousand years ago]
L3= Nubian [15 thousand years ago]

R = Proto European [11 thousand years ago]
F=Mongol [10 thousand years ago]
Z=Oriental [10 thousand years ago]
X= Proto-N. Amerindian [11 thousand years ago]
A=Adamic-Amerindian [11 thousand years ago]
B=India-Amerindian [11 thousand years ago]
C=Russo-Amerindian [11 thousand years ago]
D=Oriental-Amerindian [11 thousand years ago]

V, K= Scandinavian [6 thousand years ago]
I, W = Greek [6 thousand years ago]
H, J, T=European [6 thousand years ago]

Like the Y-Chromosome map, there are many variants of these flow maps as well. Again, it should be noted that these are my names rather than ones used by others. Please notice that there are not nearly, as many "Mutations" associated with Mitochondria DNA and the Y-Chromosome so we will concentrate of Y-DNA Haplotypings mostly.

Caution

There is a large group trying to push the E Haplotype [Y-DNA type] into all sorts of interactions, but there is little evidence of that. I will explain some of that as we go along. One of the better known population distributions concerns the expansion of Ireland and the UK. I will go into that as that population drift affects many and a substantial amount of secondary information can be used to form a more probable lineage of this group as well as others.

.Please note that the 5 American MtDNA groups don't make sense. While MtDNA shows very few mutations compared to the Y-DNA, over might the X, A,B, C, and D MtDNA mutations formed on both continents AT THE SAME TIME. It was as if some of the people were brought over to America to live in flying machines.----Forget I said that. Maybe all the different mutated people got together one day and made a vow of no-crossbreeding, walked across Asia to that stupid land bridge thing, and ate some polar bears and such as they finally got to the Americas to started breeding again. I don't care which one you believe, just know that it is odd. Speaking of odd; mitochondria themselves are as odd as you can get.

DNA Mutation

I need to give you a quick overview about DNA and the weirdness of Mitochondria. DNA is a secret part of you that tells where you came from. Instead of saying race, we now say haplotype. Simply put, a haplotype is a colony of individuals that have mutated the same way. Yeas I know that is "race" but English is changing. It seems that major changes in characterizations are sort of recorded in the DNA structure such that one can look at a sequence of parts called Alleles that are stuck onto what can be referred to as a "Genetic Segment" and determine your ancestral "Haplotype". People with similar Haplotypes are similar because their DNA is somewhat similar. This Haplotype thing not only clusters people by characterization but also when the "grouping" was formed so one can determine differences of people groups, societies, continents and how these groups moved from one place to another. I know you have seen DNA used in a courtroom to tell is someone killed another if the killer left a hair in the room, but I'm talking about using this stuff the trace humans back through time to their origins. Here is where the sex chromosome and Mitochondria come in.

Ancestry

Two different Haplotypes can be used to determine ancestry most easily. One is found in a Y-chromosome [Y-DNA] and the other is from mitochondrial DNA [MtDNA]. These Haplotypes have different designations. Haplotypes pertain to deep ancestral origins dating back many thousands of years.

144

According to "some" research, Y-DNA is passed solely from father to son, while MtDNA is passed down the maternal line to both sexes.

Mitochondria DNA

This whole mitochondria thing should be getting you confused. If not, let me enlighten and confuse. When someone talks about Mitochondrial DNA, he is NOT talking about the DNA that makes you who you are. Instead, people have a **second set of DNA inside themselves** **from aliens** called mitochondria. Mitochondria are small "quasi-animals" that lie in the cytoplasm of your cells. It is believed these were once completely separate bacteria, but over time, they became "part" of the cells and help supply energy to the nucleus [where the 23 pairs of "real" Chromosomes are hiding as shown below].

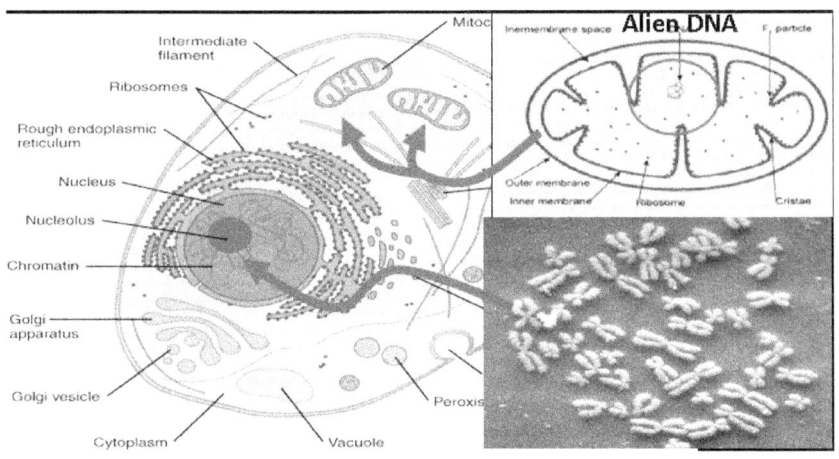

To show it is an alien, the mitochondria has its own DNA different than that of the cell that is copied from Mitochondria in the mother [most of the time]. If the mother had gone through 50 mutations since the animal or person became who she was when she carried a new baby, the mitochondrial DNA that was passed on would be an exact duplicate and carry the information of those 50 changes or mutations. All one must do is decode those 50 codes and every change ancestor and mutation would be known. How simple is that??? To make it easier, some scientist calculated that there was supposed to be 7.5 mutations every million years. **Unfortunately,** for those trying to make this simple; mutations are very randomized and clumped together during various cosmic events. A number of estimates have the mutation rate at MUCH higher rates making all the timing much more compressed than previously determined. This is where that new timing chart will be handy.

Why are Mitochondria from Females?

I suppose you are wondering just how mitochondria DNA only come from women, but the idea is that [most of the time] when sperm enters an egg, its tail, holding the sperm's mitochondria doesn't make it. Either the sperm-tail is simply amputated or when it enters, the tail dissolves so the new entity HAS to use the mother's mitochondria. Unfortunately, for those trying to make this simple, male tails are not always consumed, so some mitochondria are passed by the male. The image below shows sperm trying to get into an ovum.

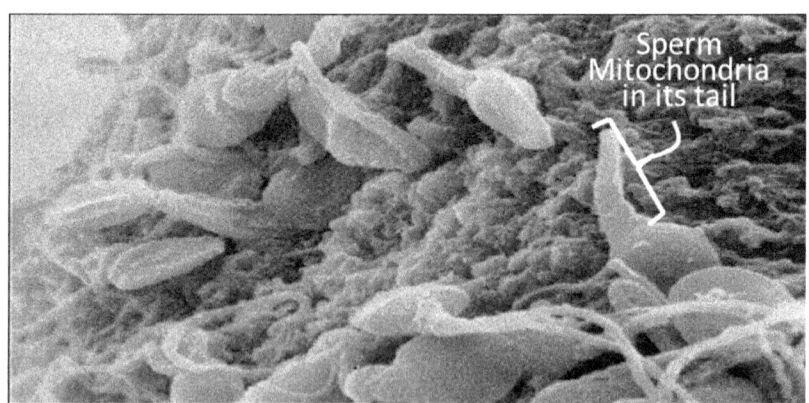

Y-DNA

Similar to the Mitochondrial DNA, Human Y-chromosomes hold mutation details of a male and most of the Y-chromosomes are not even recombined during association so the record of mutations is also held by these Y-

Chromosomes. An example might be that if there were 2 generations of identical Y-Chromosome males followed by a mutation. Two sections showing the previous structure and one section showing the mutation would be recorded on the Y-chromosome so one could tell number of generations, number of mutations, and the general time period for each.

Tale of the First Man and First Woman

For any of this to have major truth, there must be a viable community or survival would not have been possible. That being said, the first man and first woman have not been determined to have been near each other when the first inception happened. The mitochondrial EVE is much older than the Y-DNA Adam. That just makes no sense to me at all!!!!!!!. I'm thinking Eve was with someone else!!!

 The Anak and Cro Magnon scientists were breeding, manipulating, and changing animals, and people. They probably also did like we do today and modify corn DNA so a corn worm won't eat it. Or modify pigs so they get so fat they can barely walk. I not saying these people were any worse than us, but they carried out a substantial amount of animal modifications. We know, for instance that they recreated Tyrannosaurus Rex, Hydrosaurus and a few other dinosaurs during this time and the animals were here during some of the worst wars before the end of the Pleistocene. The wars had turned nuclear and radioactive, non-fossilized dinosaur soft tissues are being recovered every day.

It didn't matter much as meteors striking the earth shifted it on its axis and almost no one survived. While this was happening, those that did survive had been changed as there a more major mutations at this time than all the rest combined.

10 Major Mutations

As described previously, we find the beginnings of what we see as "races" today as they are mostly initiated at the end of the Pleistocene Age 10 thousand years ago. From the Y-DNA Haplotyping, we find 10 specific massive mutations that will define the major races of people around the world.

- C= true Negroid
- E= Eastern Nubian
- **G=Armenian Who would spawn more of Europe**
- I = Greek [who would become white Nordic groups
- N= Russian
- O= Oriental [Chinese, Pacifica, Australian.]
- K= ProtoAsian [Japheth]
- J= Hamite [Northern African]
- R= Scythian [Balkin]
- P=Proto-Amerindian [That would define all of the Americans.]

This does not mean the Anak were not still around as they could not procreate. This type of mutation had little effect on them; however, they were getting extremely ancient by this time. After all the changes, people built massive civilizations, initiated worldwide trade, build great cities and monument, and then another disaster came along.

Six Thousand Year Mutation

As I showed previously, most of the mutations occurred either 10 thousand or 5 thousand years ago. And it looks weird. Besides the horrible idea of flying machines, many Anthropologists completely ignore the worldwide flood, the shift in the Earth's axis and the complete destruction of the world 10 thousand years ago. It makes no sense to them even though there has been a dozen studies showing the event happened, historical records from around the world [over 80 thousand stories and references known], and many religious documents attesting to its certainty. I am not going to provide all 80 thousand flood stories; I just wanted you to understand that this thing DID happen and it was at the same time that the earth shifted on its axis and the whole world was flooded as ice melted, atmospheric disorganization spawned rain and huge tidal action swept across the world.   Something else happened. With the atmosphere in disorder, cosmic rays filled the skies and mutations flourished. Key elements which greatly affected DNA distribution include the following.

- *The appearance of the Cro-Magnon* or *Adamic man [40 thousand years ago]*
- *The end of the Pleistocene Era [10 thousand years ago]*
- *The worldwide flood associated with Noah [10 thousand years ago]*
- *Alien genes found in Neanderthal and preInca skulls [from 10 thousand years ago]*
- *Nuclear Events [recorded in histories, and physical evidence such as radioactive Tyrannosaurus bones] [10 thousand years ago]*

Another Destruction Period

The world was a mess and DNA "suffered", but what about the 5 thousand year old mutation disturbance? What we find is **a massive nuclear war**. While this sounds fanciful, there is a lot of proof that is not the subject of this book.

The Oklo Nuclear Reactor field was discovered in Africa. 13 reactor sites have been found so far. While dating nuclear anything is difficult, scientist know that these have been here for hundreds of thousands of years.

Hundreds of cities were destroyed and walls were vitrified by enormous heat from the weapons [6 thousand years ago.

Skeletal remains of people in the streets are still radioactive. One city is known as Mohen-jo-Daro [Mound of the dead] there were so many people strewn in the streets.

It was the worst World War [6 thousand years ago]. We are told 1/3 of the population of the entire world was lost in the war.

The real Aryan invasion [not the one described by the British] [6 thousand years ago]. The Scythio-Aryan people pushed the Dravidian –pre-Indians down to the tip of India and hundreds of cities are being found that were decimated.

Historical references indicated many people were "changed" some even became APE-Like. [6 thousand years ago]

We even have a good idea when the wars finally ended, form dozens of different group analyses. Here are a few.
3000 BC- according to Dr. B.N. Narahari Achar planetary software and astronomical references in *Mahabharata*
3066 BC- according to Dr. D. Abhyankar subtracting 38 war years of *Mahabharata*
3067 BC- from Planetary software and description in the *Raghavan*

3090 BC- Median date for Egyptian *Zep Tepi [New Beginning]*

3100 BC- according to Dr. N.S. Rajaram (astronomical statement and interpolated passages of Mahabharata)

3100 BC according to the reunification of the upper and lower Egypt

3104 BC- according to the start of the Age of Kali [Hindu beginning]

3112 BC– according to the start of the Mayan Calendar that just ended in 2012

3127 BC- according to the *Aihole Inscription* of 7th century AD Egypt)

3143 BC - according to Shri P.V. Holey Astronomical measurements of Mahabharata

3400 BC- *M*ongulala historical reference- end of the Blood Age [Brazil]

3138 BC -based on astronomy of Saptarishi Mandal the Kurukshetra War dates back to 3138 BCE.

3100 BC- Excavations in Kurukshetra yielded **iron** arrow and spearheads dated by Thermoluminence Test to this date.

Major Mutations

While the war is not the subject of this book, nuclear anything means mutation and this was bad. The Mayans indicated that people simply forgot all of their knowledge. The Bible indicates that people could no longer talk to anyone who did not know their verbal language. Our brains began to atrophy as we now could only use about 10 percent of its capacity. The y-DNA Haplotypes show 6 more major deformations of the DNA because of this war.

- **R2=** Aryan nation was created and Aryans pushed the Dravidians out of India.

- **H=** Afghan people arose and established that area

- **L**=Dravidians of India were forced to the bottom of its peninsula

- **R1a**=Slavic established the beginnings of Baltic nations.

- **R1b**=Gaelic people arose and became Europe.

- **N1b**= Scandinavian, White Nordics, moved north to become Vikings and the like.

6 thousand Years Ago

Nuclear World War Abounded. The Biblical book of Jasher tells us that 1/3 of the population of the entire world was destroyed in the war and 1/3 were so mutated that they became like animals. The remnants of bodies, still radioactive, have been found along with globs of glass that once were pottery. Melted stones on building and walls help us understand that nuclear bombardment also would cause substantial DNA Mutations. As the Wars subsided, around the world this time was proclaimed as the "New Beginning" "Zep-Tepi to the Egyptians, The new Age of Kali to the Hindu. The Pre-Maya started a brand new 5-thousand year calendar. On and on we could go. As the radiation began to subside, mutations began to subside as well. Let's look how the mutations changed how long people lived.

Lifetime Mutation

As I mentioned before, very few mutations on humans occurred before 12 thousand years ago. The reason is simple there were not very many generation of Homo-Sapien-Sapien before that time. While we have timed the beginning to about 40 thousand years ago, that would represent only 7 or 8 generations. If we read the ancient Jewish book "Adam and Eve II", it states Adam lived 5500 years and we can assume the other Adamic people before the flood lived about the same length of time. The charts following are the Jewish king lines, the Sumerian, Chaldean, and the Egyptian king lines. The

Chaldean and Egyptian timelines were compressed slightly to line up the critical events, but the slopes are the things to look at. Overnight, all societies around the world were affected by something and life-spans were truncated. First is the "normal" Adamic/Cro Magnon Race timeline.

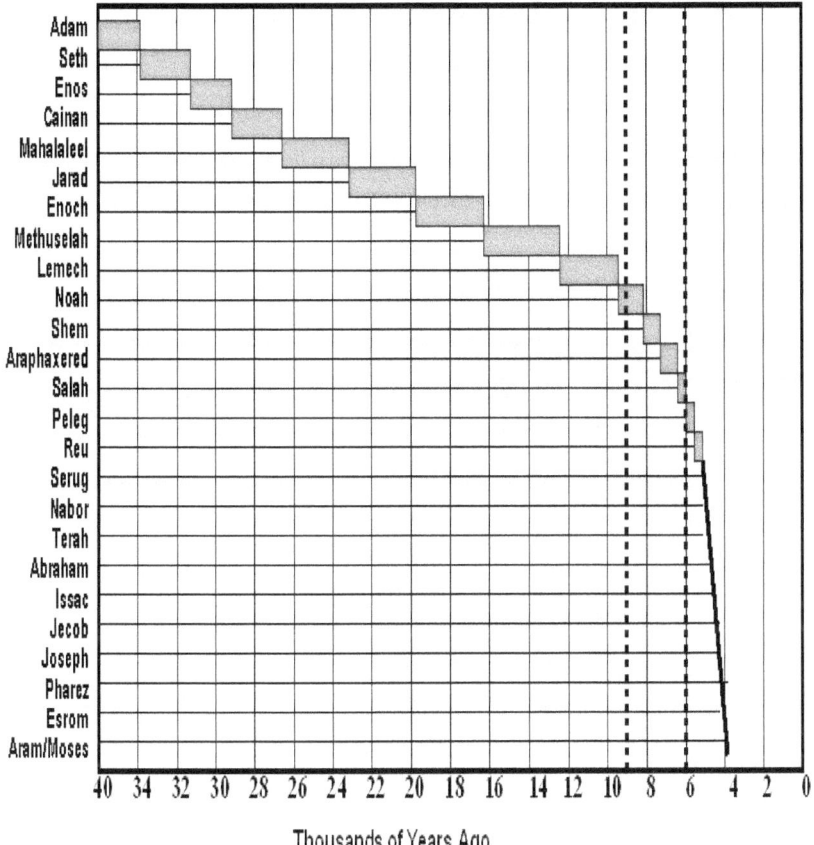

Thousands of Years Ago

Chaldean King Timelines

In the Jewish lineage, Serug was the first king after the Tower incident and his lifetime was drastically shortened to only about 200 years. The Chaldeans monarch "Tudia" had the same unusual life-spans. Six thousand years ago, all the rulers began to have "normal" life-spans.

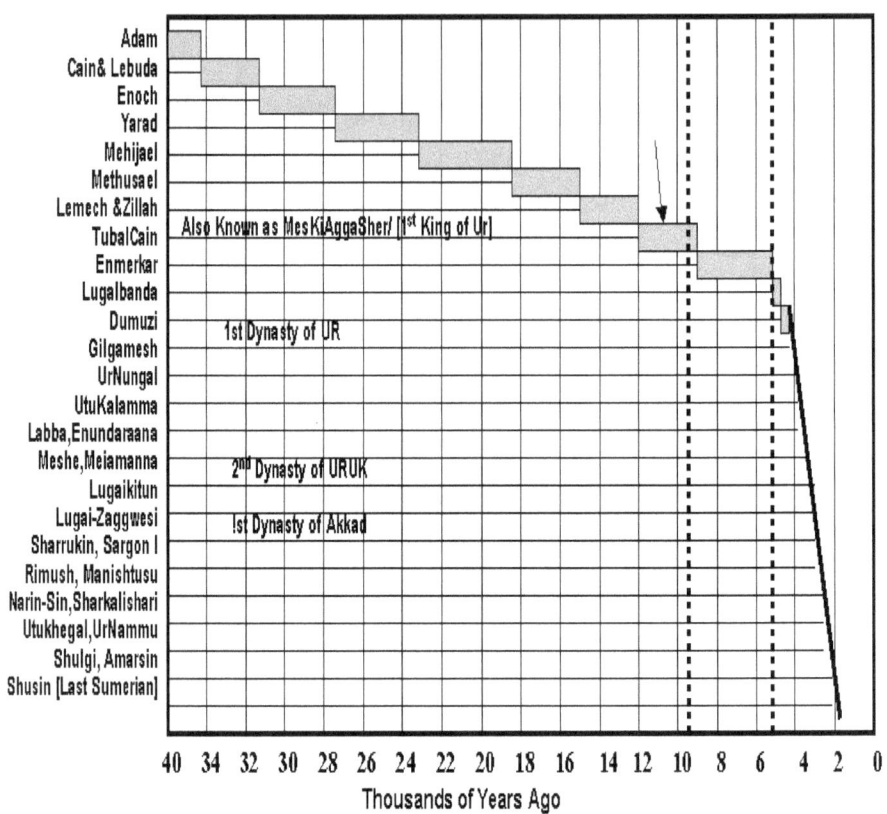

Thousands of Years Ago

Sumerian King Timelines

According to Hittite works "The Epic of Creation (Enuma Elish)" and "Epic of Gilgamesh", the Sumerians tell about the same story as the Chaldeans and the Jewish histories. This is important in that Gilgamesh and Noah may have been the same King. In the "Book of Giants" [ancient Dead Sea Scroll book] Gilgamesh was a prominent authority figure who was "the only survivor" of the flood. In the Sumerian text, there was a Gilgamesh that became king after the flood but this seems to be a different person entirely. We can be sure Tubal Cain and Noah were different flood survivors as they landed at different mountains after the flood. Given that King Inrush had been an ungodly king and King Noah of the Kingdom of Canaan was a Godly one, it could very well be that the second King Noah was the Gilgamesh described by the ancient

154

Sumerians and in the book of Giants. As the Gilgamesh story trails off after his adventure, I will start with him and the go to another Kingly line of the Sumerians.

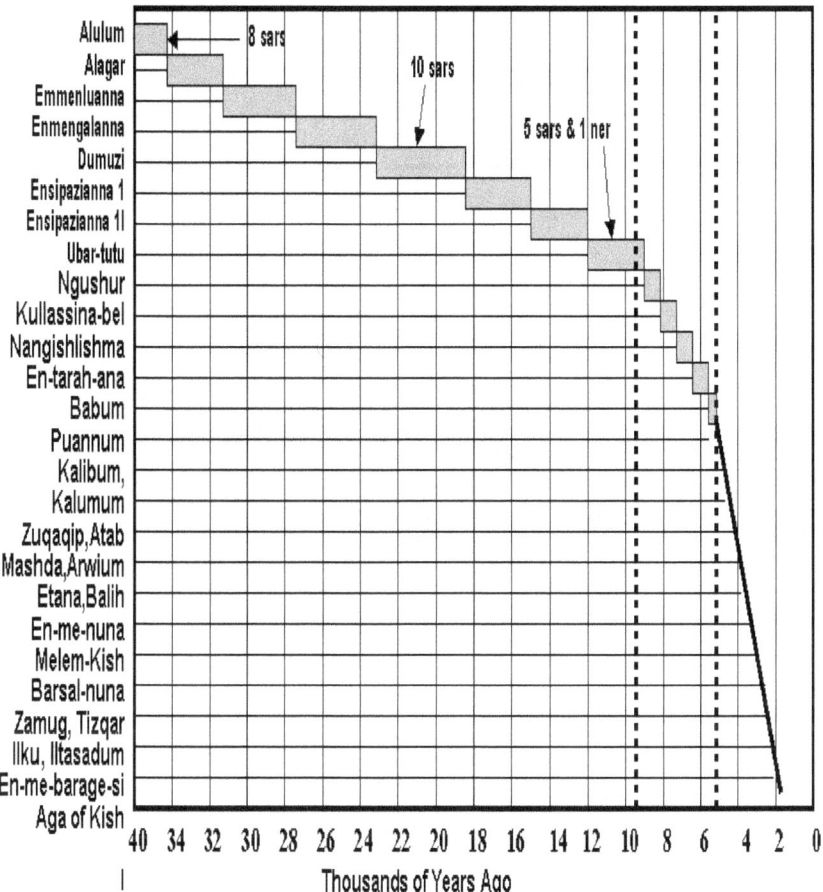

Please notice this same mutation noted after 6 thousand years ago. In every case, life spans were greatly reduced to about 150 years and even shortened a little after that time.

Greek King Line

In Greece, we find the same. While the details have been glamorized, we can certainly see the same mutation.

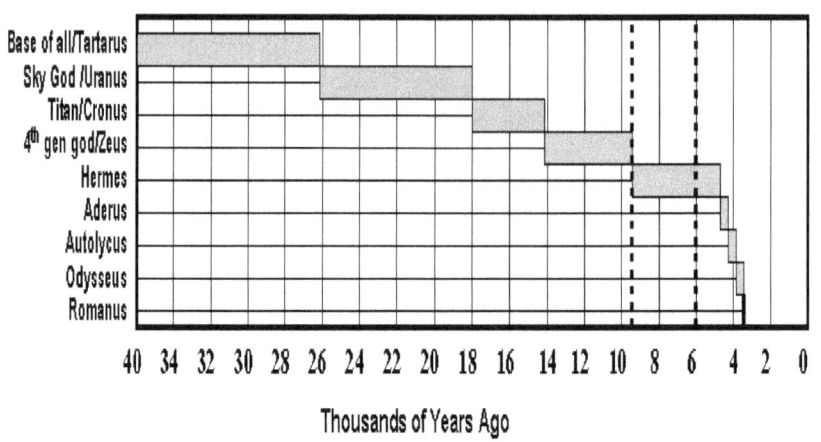

Thousands of Years Ago

Egyptian King Timelines

The Egyptian monarchs of the "0th dynasty" had the same unusual life-spans. All of a sudden, 6 thousand years ago, something happened and the ruling people began to get shorter lives. The major mutation events are depicted on each of the charts to show consistency and strangeness. The Egyptian timeline is shown next.

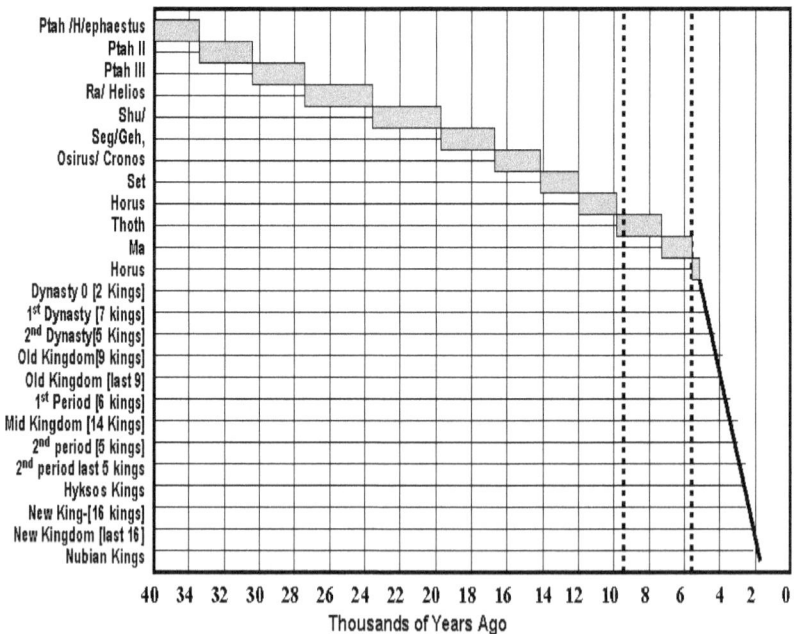

Thousands of Years Ago

I'll let you in on a secret. If people around the world, all of a sudden had shortened lifetimes, some mutation occurred. With

156

that, let's discuss some of the more "special" 6000 year old mutations. After the war, the Earth was in a mess, and mankind was in a mess. The book of Jasher tells us 1/3 of the population of the world had died in the conflicts. Massive radiations cause huge mutations and people began changing substantially. After the changes, there would be an adjoining of secondary mutations that would integrate and define the various people of today. As I mentioned for the most part, the African nations were not hit as hard as others. The "Rig Veda" tells us the Black Indian [Dravidians, were pushed to the bottom of India by the Aryans/[Scythians] exactly like the Haplotyping studies show the [R] mutation people push the [H] people almost to extinction and to the very bottom of India.

Rig Veda

The ancient "Singer" praises the god that destroyed the Dasyan and protected the Aryan color [red or red people actually interpreted "metal" color]. He is the thunderer who bestowed on his friends the fields, the sun, and the waters. The storm god who rushes like furious bulls and scatters <u>the black skin [Dravidian people]</u>. The "Sacrificer" [Indra, one of the Aryan] poured out thanks to god for "scattering the <u>slave bands of dark decent."</u> [Some have tried to force the idea that Aryan meant white, but there is no truth in it. Certainly, there was an Aryan "invasion" of sorts about 5-thousand years ago, but it was of people with the Aryan color. The reference to making the black people subservient certainly shows the black race was not the ruling race during this time as some have put forward.]

You may wonder about this red colored skin. But, look around, there is no question. Ancient rulers were red, many had long heads, and many were huge. The direct descendants of Adam were NOT. With that let me talk about apes.

Ape Race

While I'm talking about the end of this horrible war time, the book "Jasher" told us something else. He told us that a huge number of people became "like Apes and Elephants". Around the world it seems they remember the same thing. The first abnormality we will look at as proposed in the mutation chart I showed earlier is the Ape-Race. Before we get into the seemingly absurd, let's talk about chimpanzee. To discuss this horrible DNA event we need to first consider both the Chimpanzee and Bonobo. Chimpanzee and Bonobo DNA are very similar to humans. Some estimate the similarity to be 98%. That is a misnomer, but it is close. The main difference is that the 2nd set of chromosomes has been split apart as shown in the DNA sequences following. This makes the Chimpanzee have more chromosomes than a man, but characteristics are still very similar. **Today, it is believed that the Chimpanzee came after humans.** Many say while this is true, they do not believe chimps came for mankind. The normal chart is shown below as some common ancestor well after gorilla split from the homo-Sapien line. Some believe the line separates 6 thousand years ago. Unfortunately, there is a lot of evidence.

Chromosomes Show Similarity

In the Chromosome display shown next, the chromosome on the left is human and the Chromosome to the right is chimpanzee. Chromosome 2 is split in a chimpanzee, besides that, Chimpanzee looks human.

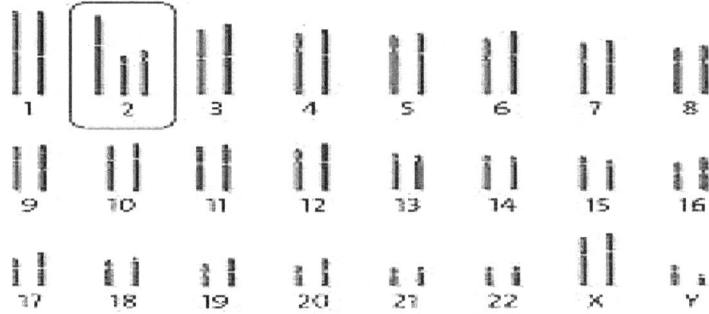

Apes After the War

This may seem odd to you but let's investigate anyway. This whole Chimpanzee thing seems to have happened when all of the major mutations were happening around the world 6 thousand years ago. The world had been at war and nuclear weapons had flattened much of the landscape. The Great Tower of Babel had been flattened. All of a sudden, chimpanzees and bonobos came into existence with similar DNA. However, the chimp genome is 10 to 12 percent larger than the human genome so we can believe it is more complex and possibly more advanced. Anyway, it shows Homo Sapiens were here before it came along. Let's see what various histories tell us.

Genesis 11:1-9 Now the whole earth had one language and few words. -- Come, let us go down, and there confuse their language, that they may not understand one another's speech. -- Therefore its name was called Ba'bel, because there the LORD confused the language of all the earth. [Notice it says that people used very little spoken words. The only way that could be possible is that most communication was done without words. After the war, many mutations meant we had to relearn speaking.]

Jasher 8:32-39- And God said -- let us descend and confuse their tongues, ---- they forgot each man his neighbor's tongue, and they could not understand to speak in one tongue. ---And the Lord smote the three divisions that were there, and he punished them according to their works and designs. Those who said, we will ascend to heaven and serve our gods, became like Apes and Elephants [Similar to Genesis, but many died and some were mutated to become like apes.]

Jubilees 10: 24-26-And they no longer understood one another's speech, -- the Lord did there confound all the language of the children of men, -- they were dispersed each according to his language and his nation. [By saying it three times the writer is pretty sure, some limitation in man's capabilities occurred 6000 years ago.]

Totonac- Mexican Tradition-After the flood, the boat finally rested and God reversed man's face and hind parts and turned him into a monkey. [This is probable an indication of man turning to a chimpanzee during or after the war.]

Mayan Tradition-During the second creation, people turned into monkeys and the world was destroyed by wind. [This is probable an indication of man turning to a chimpanzee during or after the war.]

Aztec History-*During the age of the four winds **men turned into monkeys** according to Codex "Laticano-Vatino"* [This is probable an indication of man turning to a chimpanzee during or after the war.]

Lower Congo Tradition-*"First God created man. After a huge flood, **men put their milk stick behind them and were turned into monkeys.**"* [This is probable an indication of man turning to a chimpanzee during or after the war.]

Tibetan History-*"Tibet was almost totally inundated by the flood. The **survivors had been little better than monkeys**. The god Gya sent teachers to civilize the people and they repopulated the land after the flood."* [This is probable an indication of man turning to a chimpanzee during or after the war.]

I know you are thinking this is absurd, but chimpanzees are still around and they came from somewhere. To bring home this possibility, for a period of about a thousand years after the war, Ape-men living in a community were considered normal, and respected. Here are just a few.

Egyptian Ape Race

In Egypt, the ape was raised up as a caretaker of the precious Dendera tubes and was part of their worship services. Hundreds of these honored Baboons were depicted as the most trusted guards of the land. They worshipped with the Egyptians and they protected the Egyptians before they died out a few thousand years ago as the mutation was not very viable ---luckily.

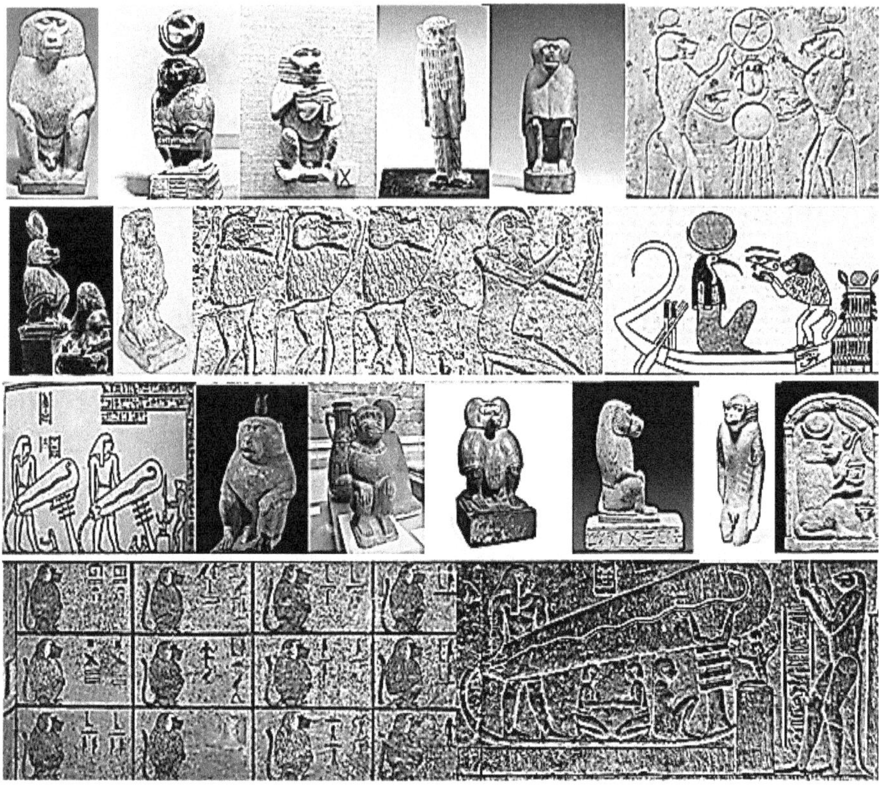

Hindu Ape Race

In Cambodia Ape guards even guarded the great Ankor Watt Temples as shown below. Why would anyone believe an ape had the mental capacity to guard Ankor Watt?

Other images in Cambodia show a great respect for the ability of the apes and monkeys of that time. Some might say they appeared to be part of society.

In India a great monkey warrior arose during the massive Babel Wars. His name was Hunaman, He is depicted below. He was the commander of an entire "monkey-man army. In India, many ape heroes abounded.

American Ape Race

In Ecuador, the Maya even worshipped some apes as shown to the right below and to the left, Aztec worshipped Quetzalcoatl the Ape god. A number of their images are shown.

  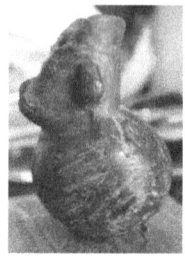 

This race of people died out in less than a thousand years after the initial mutation. Ape-men were a critical part of societies around the world. They were known for their bravery, strength, and dedication by all accounts.

Face: Simian features, large nose with forward nostrils.

Body: usually powerful body, often depicted larger than "normal" people.

Tail: Usually depicted with a tail

Soon, they were gone. [About 35 hundred years ago]. That brings us to the group Jasher called Elephants. To me they looked more like Lizards.

Lizard Race

Known as **Homo-Lacertilian**. This might be the strangest and saddest of all the mutations in DNA during this time. The Book of Jasher called these unfortunates as **Elephant People**, but most, refer to them as Lizard men. Around the world, we find evidence of lizard people becoming part of society just like the Ape people of that time. Hundreds of images were generated as lizard people became critical to the civilizations of that time. Like the Ape people, their time on the earth was limited; however, there are a number of images that show these people were able to procreate for a time.

Iraq

Iran

China

S. America

Polynesia

India

Mexico

The preceding images from Iraq, Iran, India, Mexico, Polynesia, South America, and China show these people were found around the world. If that is not enough; below are some more from Kosovo, Uruk, Yugoslavia, Pakistan, Australia, and England. Please note that the images from England show the lizard people were taller than "normal" people of that time.

Face: Long with protruding snout

Eyes" Coffee bean shaped and large

Uniform: Many of these people seem to have a number of bumps on one shoulder. The reason is not known.

Body: long slender with wide shoulders in most cases

Like the Ape-People it is believed that the Lizard People died out within a thousand years of the mutation.

American Race Controversy

The American Indians supposedly traveled from Asia, but there have always been issues. As MtDNA shows this group descended from ancient Jews, Egyptians, Greeks, and Middle Easterners directly [mutations A, B, C, and D] rather than going through the mutation cycles of the "Normal Asians". James Adair, an 18th century settler who traded with Native Americans for 40 years, wrote that *their language, customs, and social structures were similar to those of the Israelites.* The DNA from a 24,000-year-old corpse in Siberia was analyzed. American DNA showed no resemblance to Asian populations, only to European, yet it showed a clear connection to Native Americans.

The Y-DNA has the same anomaly. Above are sample 'Q' Haplotype people from Peru, Bolivia, Ecuador, Oklahoma, and Alaska. They are essentially, the same group. When people read these Haplotypings, they don't like what they see. For instance, the stupid idea about the little bridge crossing between Russia and Alaska doesn't make sense? If people did that, one would see the DNA trail of that infusion. Look at the next flow map. In this case, we are only looking at the Haplotype called Q which makes up most of the American people. The darker the area, the higher the percentage of Q type DNA there is. Notice that Russia has almost "NO" Q. I was going to say if the Glove does not fit, you must acquit, but I think the better thing to say is the land bridge thing is bogus. It should be noted that the eastern border of North America has a substantial amount of R1b [Gaelic DNA just like Ireland]. The Q DNA descends directly from either P or R about 8 to 10 thousand years ago so we need to look further.

Somehow, America was settled **without going through Asia**, yet, most scientists kept on telling us they had to go through Asia. In fact, this farce is still being taught today so that no one would have to explain how 10 thousand years old settlers came to the Americas without walking. By having this bridge thing, no one had to come over the Ocean in a flying ship, but that is another story. As the DNA proves, there was a way for early people to get to America without going through Asia. Instead of walking, they took an airplane. I'm sorry it that is hard to believe, but researchers have found artifacts, images that can only be seen from the air, massive numbers of documents and carved images of these flying machines. Without them, many things do not add up.

Face: rounded

Skull: Brachycephalic

Eyes: round or slanted [depending on oriental influence]

Nose: Broad, but not as much as negroid

Skin: Brown to reddish brown

Hair: Black and straight

Build: Less slender than European cousins.

Clearly following the mutations that occurred after the Pleistocene extinction, people with P and R Haplotyping mutations were "Shipped" over and populated the region without going through Asia. As mentioned before the long headed Anak People were already in possession of the land and would have become the rulers of these people until they died out about 3500 years ago.

The Other American Race

Yes, I purposefully missed the other American Indian Race, we think of with a tomahawk in one hand and a bow and arrow in the other. Please notice that these people look like the reconstructed Scythians I showed before [below right].

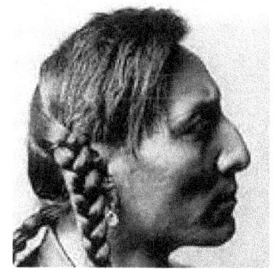 

As I mentioned before the Scythians are noted as being the Aryan race that almost destroyed the Dravidians in India. Let me tell you why they were called Aryan [metal men]. There are 2 possibilities. Either they were the reddish color of copper or the reddish color of rusted Iron. If the Scythians were brought to America in flying machines [sorry for saying that again], they would spawn a race of reddish skinned "noble nosed, chiseled faced, strong chin, individuals with a desire to be warriors. I'll let you connect the dots as we go to a very special modification of the Greek {I} mutation. This group would eventually control the Northern countries.

The ANAK people who established the Americas were already over in the Americas. Certainly, they brought "normal" people with them from (A/ B/C/D)f and Qm Haplotypes, but they didn't have an outrigger paddling the Pacific and no one was starving trying to take that northern route. They went by air except for the much later R1b Y-DNA and X MtDNA groups that possibly traveled by boat across the Atlantic. Over in Europe things seemed to be easier to understand.

Siberian Race

This group [N mutation] stayed in the coldest regions of Asia. They became the Siberians.

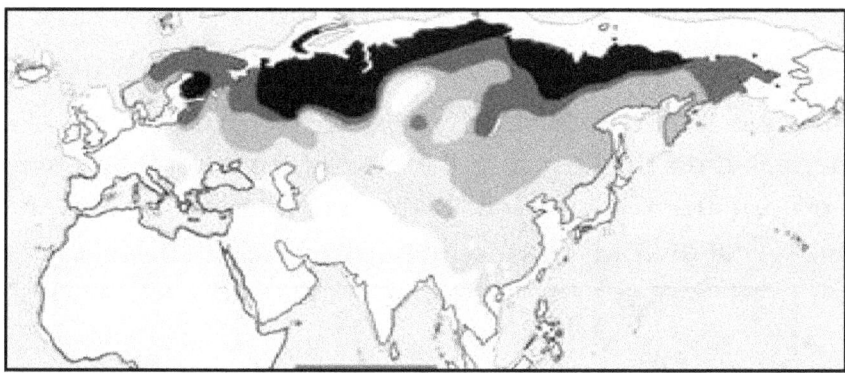

One would expect that if a massive group of people went across Asia to the Americas, they would pick up some Haplotype "N" traits. "N" genomes COVER the north Asian continent, but here is the rub!!! They are not in the Americas at all. The darker the color the higher the percentage of "N" mutated people.

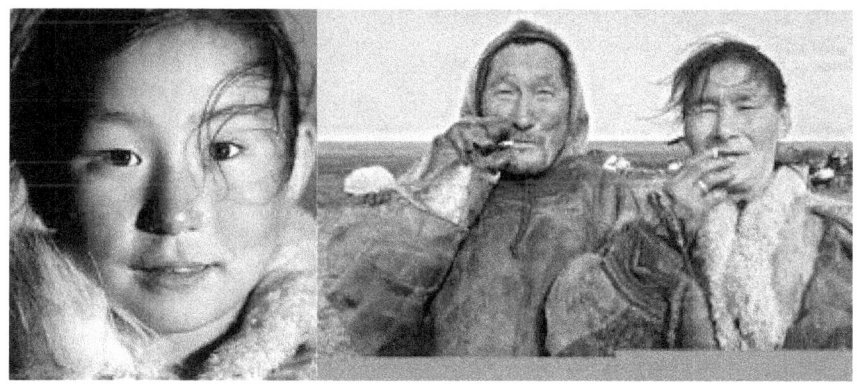

Mongoloid Race

Still trying to stay isolated, the Oriental "O" Haplotype shows up as one would expect. If the people who walked to the Americas didn't come from the Northern regions, they would have been from Haplotype O that completely took over the Orient. That didn't happen either.

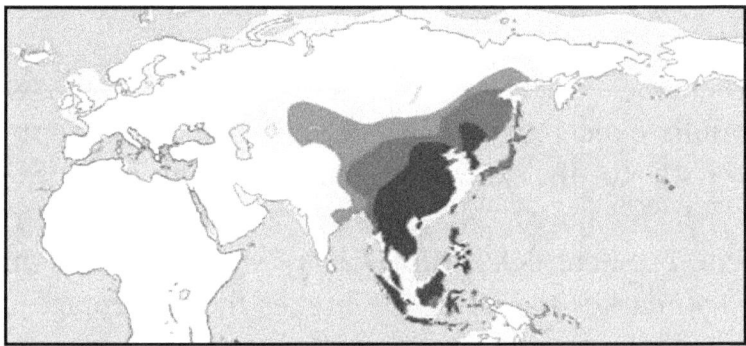

Here are some examples of this unique group.

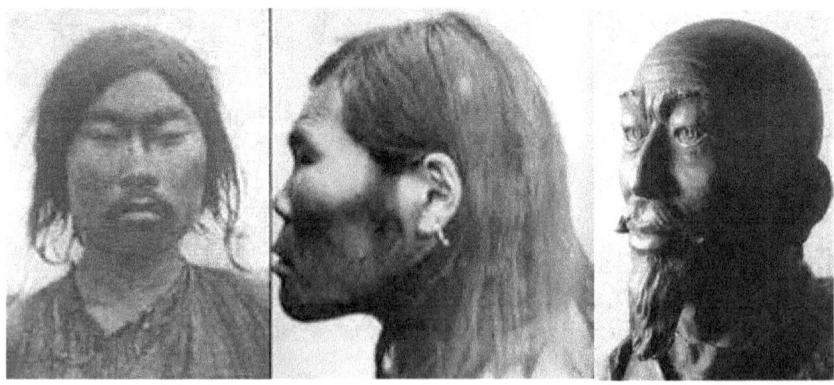

Bone: *Mongoloid subjects had about "20% higher bone density at the angle of the mandible" when compared to Caucasoid subjects.*

Skull: *mesocranic skull, fairly large and protruding cheekbones, nasal bones that are flat and broad, a nasal bridge that is slightly concave without depression*

Eye: *supraorbital ridges low frontals, marked post-orbital constriction, prominent and protruding occipitals,*

Complexion: *yellowish skin color*

Face: *comparatively flat faces, broader skull, broader face and flatter roof of the nose.*

Body Hair: *Mongoloid males have "little or no facial or body hair".*

Hair: *Mongoloid hair is coarse, straight, blue-black and weighs the most out of the races.*

Sweat: Mongoloids have 450 sweat glands per square inch while both "American blacks" and Caucasians have 750 sweat glands per square inch.

Generally speaking the people stating in the Middle East went north to become Europeans.

White Nordic Race [U/K: I]

This is an important race. The Northern European White Nordic Race is the <u>most abundant racial contribution among the primitive Western Hunter-Gatherers described by modern population genetics</u>. A typical White Nordic is shown below. His Y-DNA mutation base is the I group while the mtDNA mutation is from both the "U" and "K" groupings. What that really means is that as the Middle Eastern Semitic and Armenian races pushed north to Greece, They saw some good looking "U and "K" based girls and the rest is history.

Skin: *his skin is neither pale, nor milky-white, nor rosy or ruddy. It is rather "golden", in harmony with the hair, and seems suitable to get a moderate tan without getting burned when exposed to sunlight.*

Forehead: *is high but not completely vertical.*

Psychologically: *this is a noble, harmonious, serene, serious, patient, well-balanced, martial, honorable, disciplined, hones, efficient and racist race, but also somewhat naïve, too angelic and not very cunning in many ways.*

Stature: *Very tall*

Physical constitution: *Slender, athletic. Well-shaped, broad and straight shoulders. Long neck. Although it is a "thin" physical type, tends to develop musculature under proper conditions of diet and exercise.*

Eyes: *Ice-grey, very light, almost whitish. Sky-blue eyes are mixes between the light-grey shade of White Nordics and the dark blue of Red Nordics. The grey eye color is most common in Finland, the Baltic countries, Belarus and the European part of Russia. Large pupils, short-medium distance between eyes.*

Eye Shape: *He has an elongated eye shape. Eyes deeply inserted in face under eyebrows that are low, narrow, moderately bushy and bring a thoughtful and audacious expression and a penetrating, aquiline and intense stare. Small eye sockets.*

Nose: *Narrow, straight, not very fleshy, harmonious. The key of the White Nordic nose is that its "root" lies very high, almost in the forehead, so that the space between the eyes does not look deeply-set like that of Red Nordics. The White Nordic nose corresponds to the **well-known "Greek profile"** of classic statues.*

Ears: *Thin and elongated*

Mouth: *Thin and dark lips, with a clearly "sketched" outline. The infra-nasal depression is broad and clearly marked, in such a way that the central tips of the upper lip look separated. He seems to have a slight "fed-up" expression*

Teeth: *Lined-up set of teeth, hardly much difference between the shapes and heights of each tooth.*

Hair: *Platinum blonde, almost white, straight, thin and lank. When it grows, it tends to stick to the head.*

Body: *Skull: Dolicocephalic (long seen from the side, little width, narrow temples). Curvoccipitaly (highly convex occipital and parietal bones). This race has developed cranial capacity backwards and forwards.Face profile: Not totally vertical but almost. Straight and progressive.* ***Forehead:*** *Straight, broad and almost vertical.*

Jaw & chin*: Harmonious and well-developed jaw. The chin is between the prominent and massive type of the Red Nordics and the retracted one of the Armenians, but closer to the "RN" y-DNA model.*

Youthfulness: *a very youthful appearance, maintained until a very advanced age (although not as much as the Red Nordics).*

Attractiveness: *Abundance of athletic and active women, attractive and of great beauty, which have resulted in a very high reproductive success of White Nordics*

Paternal lineages (Y-DNA): Clearly Cro-Magnon with mutations from Armenian and Greek "group I and I1".Approximate distribution of **paternal I lineages** in Europe shown below.

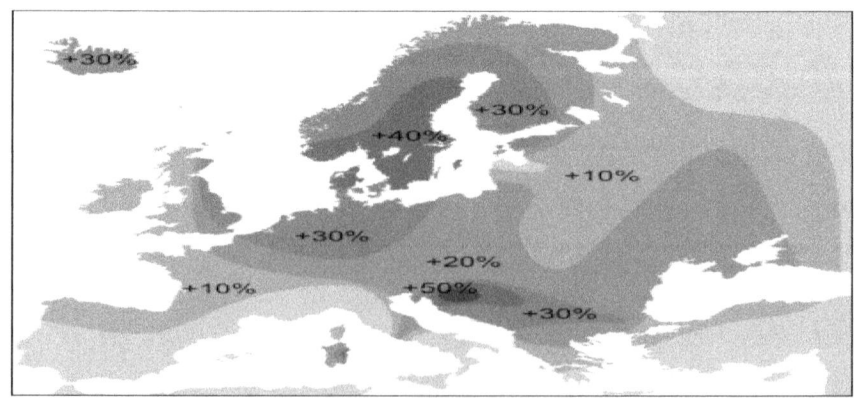

Haplotype I1 distribution (corresponding to Scandinavian types like the Vikings or the Normans) shown next notice major concentration in Scandinavia.

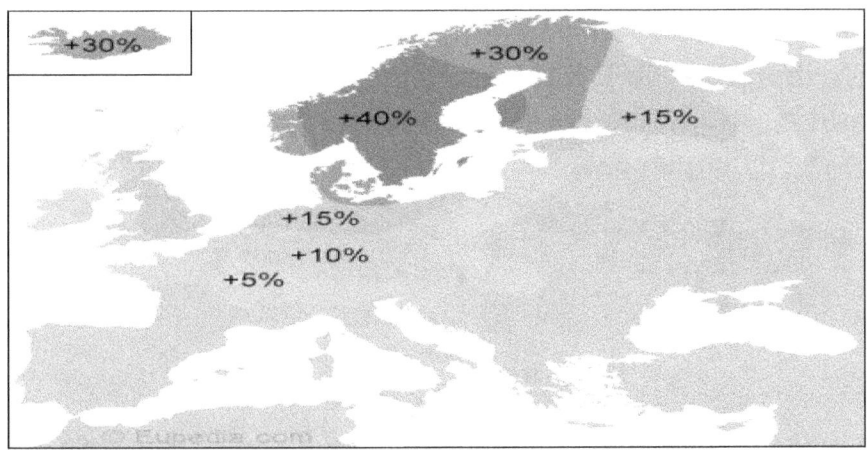

Maternal lineages (mtDNA): U and K mutations- Not all U sub-lineages are White Nordic in origin. They come from Armenia. K maternal lineage distribution in Europe is shown below with France, United Kingdom, and Germany having the highest percentage of K mutations.

Psychology: Love for honor, respect for authority. Intelligence, thoughtfulness. Highly developed willpower, <u>unable to cheat</u>, so they are useless in diplomacy. This race is not shrewd.

Distribution: The White Nordic race has expanded around the world mostly in the northern latitudes. As shown next; darker shows higher percentages.

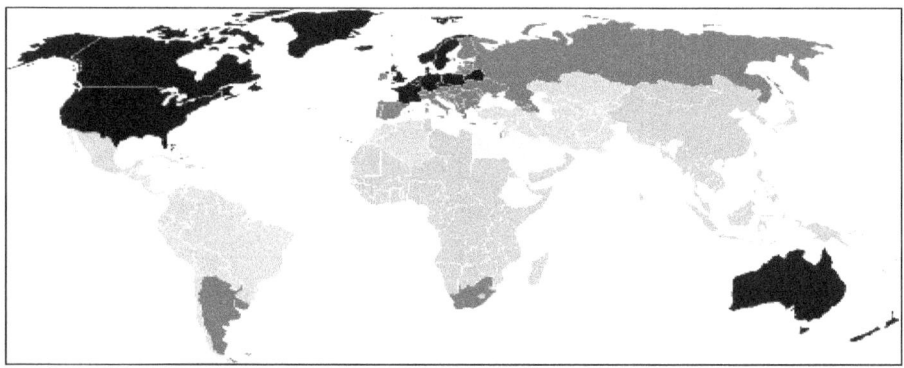

Brief history: Haplotypes genetics indicates that White Nordics associated with I Haplotypes mutated and therefore evolved from Armenian ancestors. The IJ lineage (that appeared during the Pleistocene, split in two parts, giving rise in the Near East to the J Haplotypes ("Semitic"), and in Europe to I Haplotypes. Some have the Semitic J as the base with the Armenian mutation.

Present context: Among all original human races, the WN is the one that has at present larger quantities of completely pure individuals, partly because it is probably the youngest race, and partly because historically it has displayed more racism than other races. That brings us to the light skinned Red people.

Red Nordic Race [R:H/V]

The Red Nordic race is the most abundant racial contribution among the primitive Ancient North Eurasians described by modern population genetics. That being said, Let's see what we have. The image below shows this mutation we call Red Nordic.

Squared features: *prone to gain body mass. Notice a harder, broader, more robust physique.*

Blotchy *facial complexion.*

Fighters*: red Nordics inspire restlessness, impulsiveness, brutality, aggressiveness and "explosive force", as well as a higher tendency to tyranny and "abruptness". It is also a more passionate, stronger character and temperament, as well as more muscular strength.*

Eyes: *Wide. Dark navy blue, small pupils, middle distance between eyes. Big sockets. Straight, horizontal, very sparse and almost white eyebrows.*

Nose: *Shorter, wide, rounded fleshy tip, that its "root" is not located between the eyebrows, but lower, so that it is "deep set" between the eyes.*

Ears: *Thinner and smaller than the White Nordic ones.*

Mouth: *Extremely thin and narrow lips. The outline of the lips is not clearly defined, nor differenced from the rest of the skin. Big mouth.*

Teeth: *Lined-up set of teeth, smaller differences in height and shape than in white Nordics. When mixed with Armenians, the* **separation between teeth** *tends to increase. This can be due to the heredity of a big, spacious mouth which cannot be fully "filled"*

Hair: *Orange, straight. Tends to stand up instead of falling.*

Body hair: *Medium-scarce, bushy sideburns and goatee.*

Skin: *they're unable to produce melanin. Red, rosy and bloody skin. In mixes with other races, the bloody appearance tends to retreat to the face, and in the face, to cheeks, ears and under the eyes. People with a* **tendency to blush.** *Higher risk of skin diseases when exposed to the Sun more than they can tolerate, which is not much.*

Skull: *Brachycephalic, flattened occipital bone, but prominent temporal bones. Larger cranial capacity and broader face than the white Nordic.*

Forehead: *Very high, straight and vertical.*

Jaw & chin: *Strong, squared, broad and robust jaw. Prominent and sharp chin, which seems to end in a fleshy ball.*

Other features: *Accelerated metabolism, very active blood circulation.*

Youthfulness: *very youthful look, whose freshness is even better preserved than in White Nordics. Tendency of perspiration and sweating.*

Higher sensitivity to pain*: Other studies suggest higher sensitivity to thermal pain and lower sensitivity to electric pain.*

Higher lactose and alcohol tolerance *than any other race*

Stature*: Medium-low.*

Paternal lineages (Y-DNA): R (R1a, related to Slavs and **Aryans**, and R1b, the predominant Haplotype in Western Europe). R2 is also probably Red Nordic in origin. Distribution of the **R1a lineage**, the most common in Eastern European countries is shown next. It is also clearly related to the Aryan invasions as I mentioned before. .

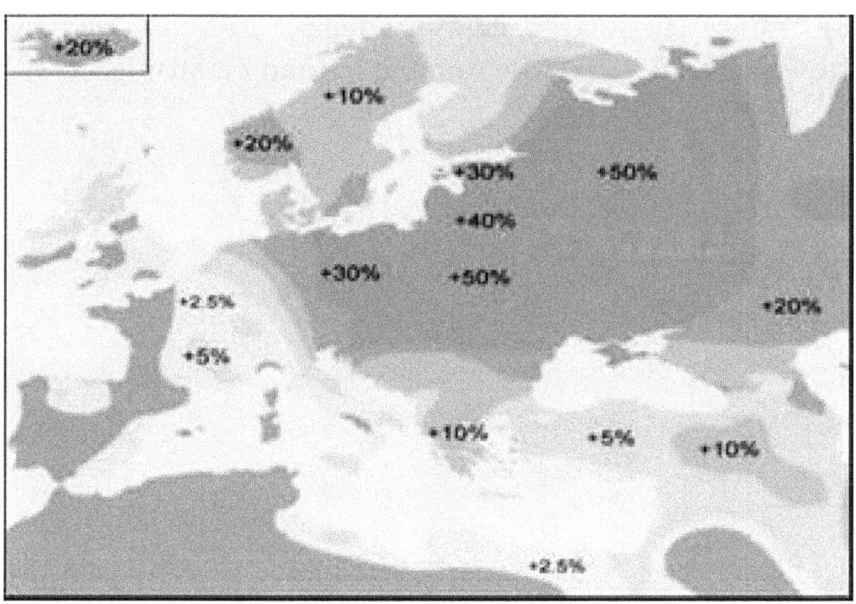

Maternal lineages (mtDNA): H, V. As shown below, probably not all HV descendants are Red Nordic, many of them can be the more ancient Armenian.

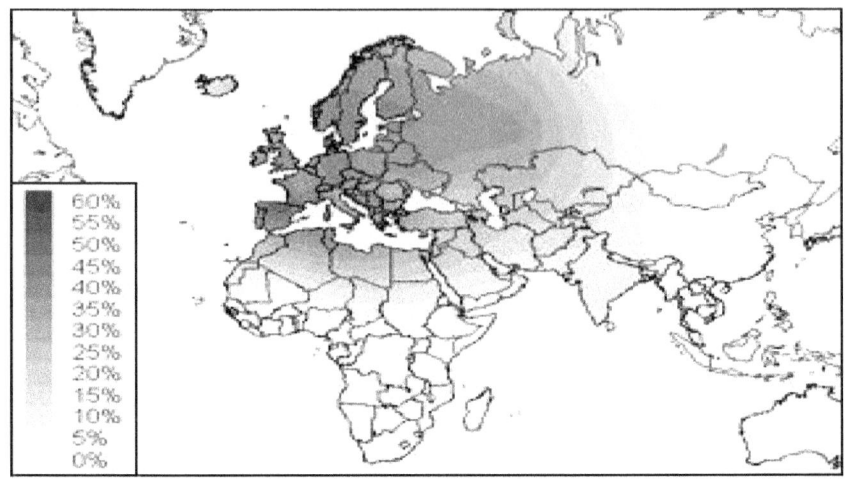

Approximate maternal H lineage distribution. Radiating from Spain, this Haplotype gets gradually lost as we move southwards and eastwards from the Iberian Peninsula.

The R Haplotype

R1a

From the Armenian descended P Haplotype, the Q and R Haplotypes derived. As shown in the initial Haplotype maps these groups would have started around Siberia-Central Asia. Some say that, from there, the Q headed eastwards until they crossed the Bering Strait and colonized America. The R ones headed south and westwards and gave rise to the R1 and R2 Haplotypes around the Afghanistan area. R2 Scythians pushed south and remained in India, while R1 split into R1a lineage towards Southern Russia and the R1b went towards Europe. R1a spread from Russian into Eastern Europe. During their expansion, the R1a societies will expand through the Middle East and other regions of Asia. R1a is associated with Slavic, Aryan and Indo-Aryan populations. These guys went all over as shown below.

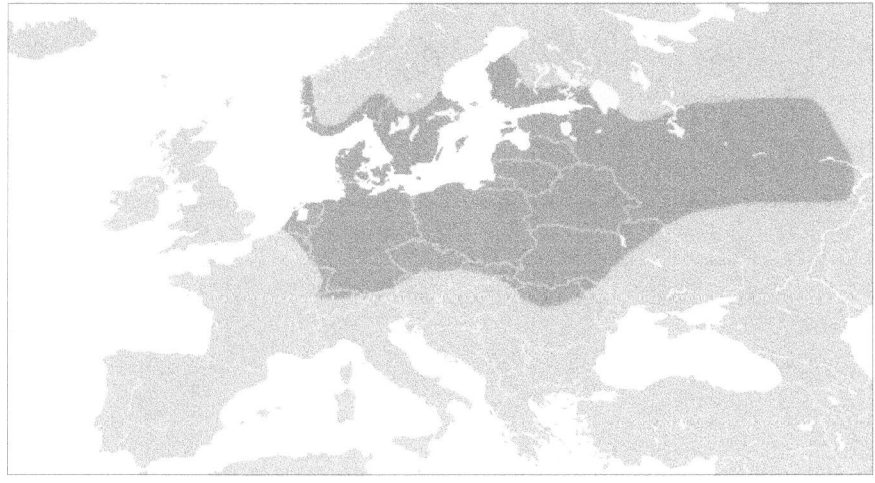

R1b

R1b's history is rather more complex. After its long journey across the Middle East, R1b people crossed the Caucasus and entered Eastern Europe near the Danube. The Bell Beaker culture is probably one of the first manifestations of the RN race in Europe. The former hunting territories of the Cro-Magnon communities were deserted: the White Nordic race, identified with Cro-Magnon man, had moved north. R1b, thus, occupied the area from France to Central European. They even became the Halstatt culture from which the Celts arose. Finally, all Western Europe became R1b people. The chart below provides a quick overview of how R1b may have taken control of Europe.

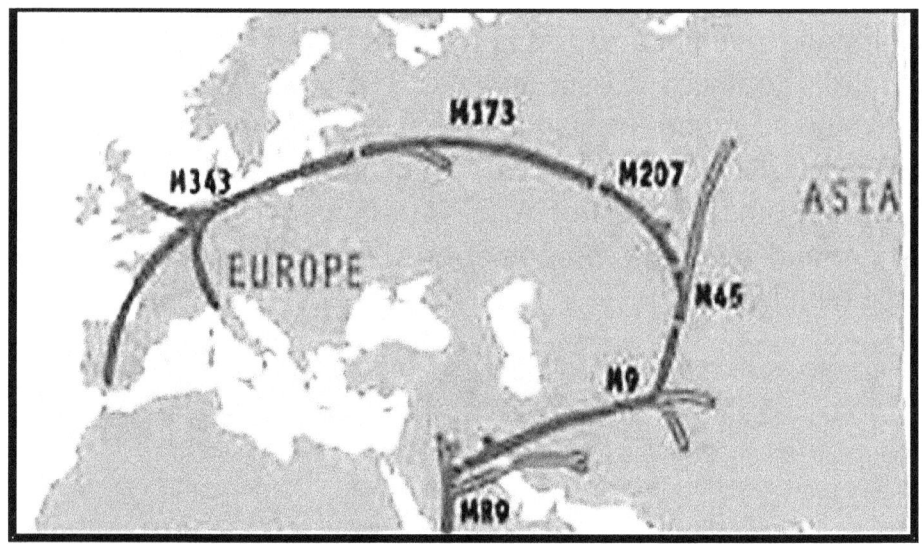

Armenian Race [G]

Near Eastern Armenian race is the most abundant racial contribution among the primitive Early European Farmers described by modern population genetics. Typical characteristics are shown below.

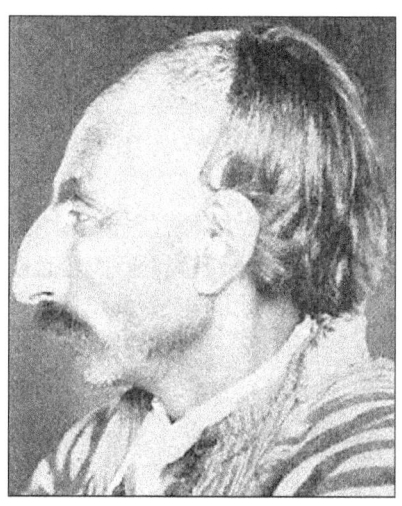 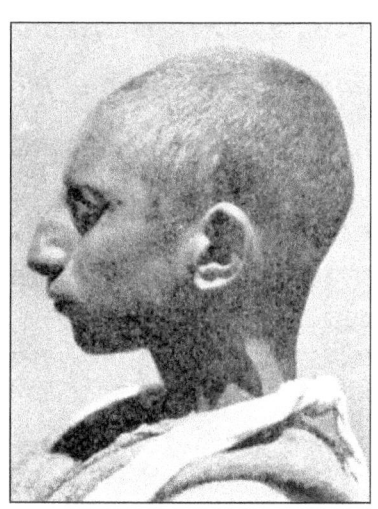

This individual is almost pure Armenian from Syria. In Europe, such pure types probably no longer exist.

Profile: *Wedged profile shape, receding forehead,*

Nose: *huge nose with a very high nasal bridge (almost at the eyebrow's height),*

Jaw: *weak jaw, receding chin,*

Stature: *Medium-low.*

Constitution: *Slim, gracile. Resistant to scarcity. extremely fibrous, lean and vascular.*

Eyes: Brown, medium-sized pupil, short distance between eyes.

Ears: Large. More elongated and rounded than either Nordic varieties, tendency to stick out.

Teeth: Small mouth, narrow set of teeth and irregular outline (differences in heights and shapes of each tooth).

Hair: Black, thick, bushy, and straight-wavy, probably quite greasy. Tendency to receding hairlines in the top corners of the forehead.

Body hair: Very bushy moustache.

Skin: Light brown. Clean and uniform.

Head: Dolichocephalic, high cranial vault

Paternal lineages (Y-DNA): F/K Armenian origins based on paternal lineages seem to come from the "F" mutation associated with Semitic or Jewish heritage. Armenian is a strongly "K" mutation group. The distribution of the "F" mutation is shown below.

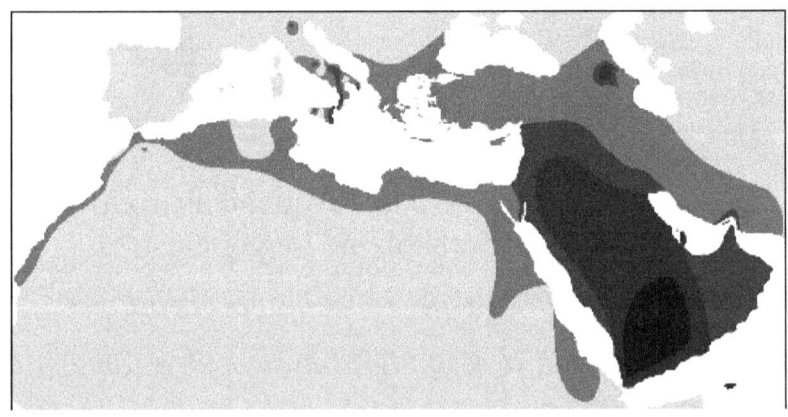

Many distribution maps have Armenians as more J and K mutations. K. Distribution of the "K" mutation is shown next. An offshoot of "F" Semitic, "K" Armenian to the P proto-European becoming R the basis of all European nations.

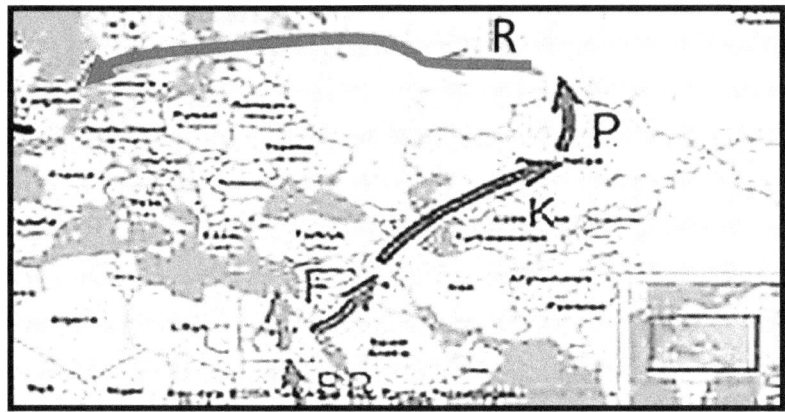

That being said; look at the distribution of the mutation called G. Notice that the almost black area in Armenia alone and no other location. This is a sign that the G mutation is Armenian alone.

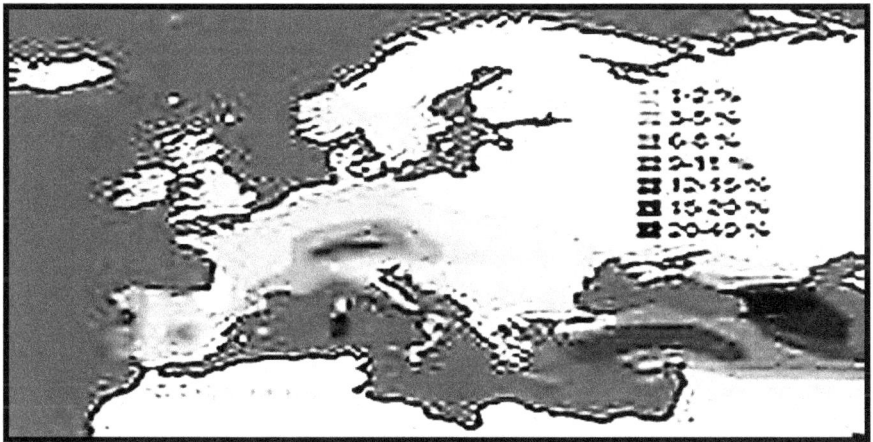

So what about the J-mutation? As I indicated before J is a Semitic group we believed was headed by "Ham" son of Noah.

Egyptian Long Head Race

This race isn't around any longer, but at one time it was an important one. We remember form Biblical history that Ham and his descendants went into Africa to a country called Egypt. If we could see some special characteristic, we might be able to develop information about that race of people. There is one thing. Pharaohs, their wives and children after the Hyksos were expelled and the Jews Exodus-ated all had long heads and long necks. Forget all the wrinkles in the mummy faces and notice that Egyptian Pharaohs ThutMoses III, Akhenaton, Tutenkomen are shown below in order.

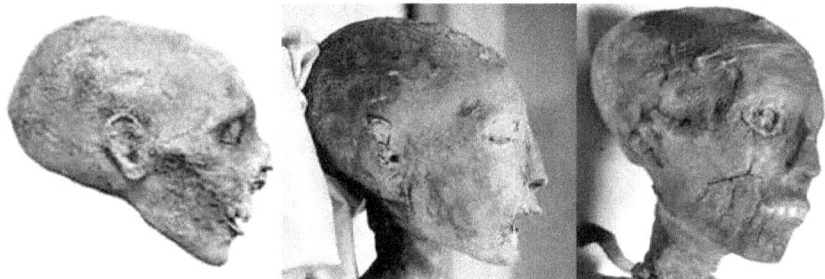

Ramesses II and Seti II [shown in order] somehow had long heads just like the first three.

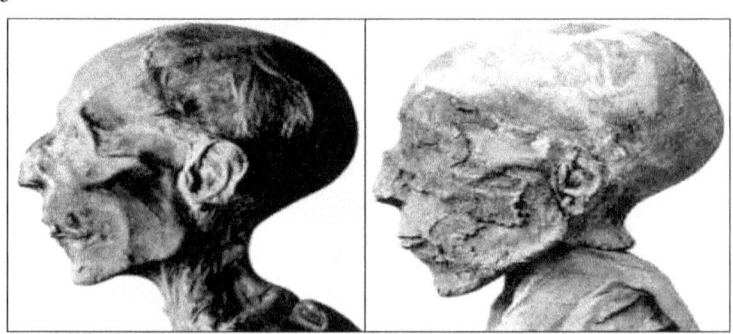

Despite having other racial influences (for instance, Ramesses and most of the others had reddish hair). While they all seem to have the typical elevated Armenian long looking neck they were more Nordic than Armenian. Let's view the kings from the front. Below are ThutMoses I, ThutMoses II, ThutMoses III, ThutMoses IV, Sebekemsaf I, Queen Hatshepsut, and Yuya [father of Tiy wife of ThutMoses III]. Here is what we see. A square head, strong square chin, broad straight forehead, slightly curly, reddish hair, straight nose starting above the eye, thin lips, and small ears. I can't tell much about the eyes as these mummy's all have their eyes shut, but they are round for sure.

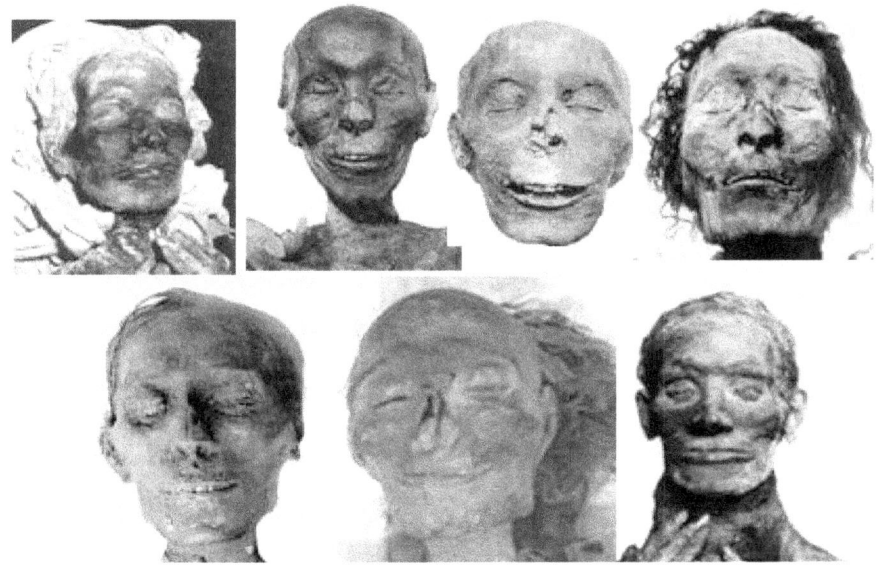

The women including Hatshepsut were all true beauties. Following are Princess Sitamun, Queen Titisheri, Meritaten and her sister Ankhesenamun. If you read studies about how Ireland got populated you would find that Meritanten is the Egyptian Princess believed to be the one who conquered Ireland for the Gauls. She was the oldest daughter of Nefertiti and Akhenaton. My book" Scythians Conquered Ireland" will give you more details about that if you are interested.

Just like the boys, the girls had this long neck long head thing. You can see that Queen Titisheri had a long head. An interesting note; Titisheri [meaning small breasted] was the mother of KaMoses and AhMoses the 2 men who ran the Hyksos out of Egypt after the Jewish Moses drowned the last great king, Apepi. Anyway; look at the next 2 side images of both Meritaten and her mother Nefertiti. Both have that long head just like the Anak that I mentioned at the beginning of the book.

Now for the interesting part; when Ham came into Egypt, he did not cross the Nile. I suppose you are wondering how I would know such a thing so I need to show you the Haplotype mapping of Hams travels. It is shown next. What you see is VERY strange. All along the East side of the Nile, the J mutation is present, but not on the other side. This is where I get a little weird in that the Anak were not able to procreate directly and we can assume the Anak controlled Egypt.

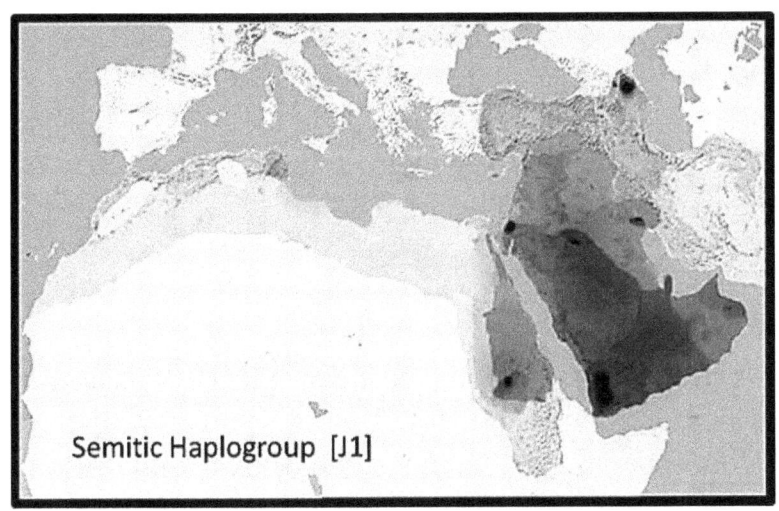

Semitic Haplogroup [J1]

OK! It sounds like I'm veering away from races of men, but stay with me a little as we look at the first King of Egypt after the horrible Babel war 5 thousand years ago. His name was Narmar and there is evidence that he controlled both Egypt and Sumeria for a time. If we find he was and Anak, it might explain why Ham's descendants stayed on the other side of the River and it might help us understand why all the Pharaohs of the 18th, 19th, and 20th dynasties had long necks and long heads. I'm not going to try to prove something here as it would mean the people of Ireland were partially Anak by way of Meritanten and I don't want anyone to get mad, but let me just look at Narmer and Naram for a minute. It might suggest the Irish people have a limited amount of Anak blood.

Narmar Ruled Egypt

For this, we find a stone image of Narmar. He was also known as Narmer or Menes, but I will show you why the Narmar name fits. The image on the stone shows him fighting the people of Egypt and dinosaurs while a flying ship flew overhead. Some think this is odd. While he doesn't have a long neck, there is no question he had a long head and he was big. REALLY BIG

On the front side of the palette, just under the king's name, is a scene depicting Narmar wearing the Red Crown that shows off his long head. He holds a mace in his left hand, while in his

right he holds a type of flail. A procession is approaching, on the right of the scene, ten decapitated corpses lie on the ground with their heads tossed between their legs. On the other side we see a similar scene. Narmar is shown with a long head again and he is big again and he has an eagle friend.

 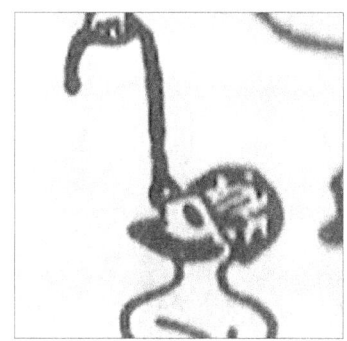

The image of the eagle yanking on the nose of a guy with a long beard seems odd as well [above left], but not discussed in this history. I found a similar scene on something from Sumeria. In a similar Sumerian tablet about "a different Narmar" or should I say Naram; this same characteristic is depicted, showing similar war technique between the Egyptian Narmar and the Sumerian Naram. I know that is just coincidence as many might yank noses.

Back to Egyptian Narmar

On the Egyptian stone there are 2 long necked. Just you run of the mill dinosaurs so let's go on. There is no doubt that Narmar was depicted as a huge ruler and some type of flying object was depicted as shown below right.

 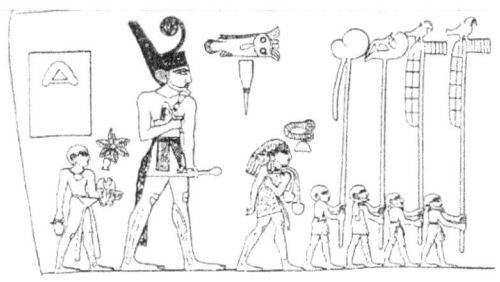

The harnessing of the dinosaurs by 2 guys wearing kilts is shown in more detail above left. Another thing of note is 2 people with horns on the top of the palette [see next]. I know this stuff doesn't make sense, but carving all this stuff in rock was not done by accident. I'll revisit the horns and the dinosaurs later.

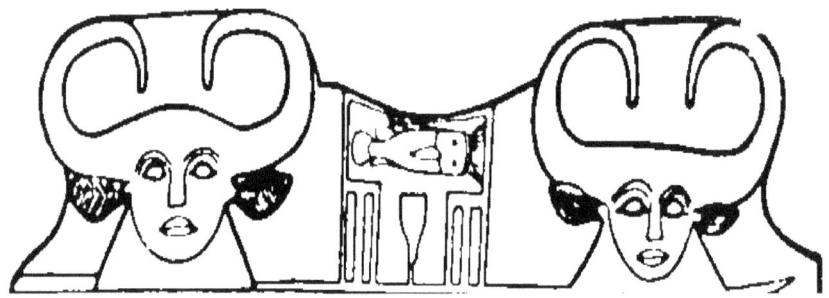

Here is one additional strange one. The thing flying in front of Narmar looks like a "flying fish" with flames coming out of the mouth area and an indication that it was moving straight upward. To make sure everyone saw it, the same "craft" is depicted twice on the palette. I'm not saying this is a flying fish, but it seems there were more war machines that simple knives and arrows.

So the giant Narmar takes control of Egypt with soldiers having long heads, with flying fish, with battles against dinosaurs, while people with horns were looking on. If that doesn't sound bizarre, I don't know what does. No wonder he beat the bad guys. If we look at the Sumerians, we find out something strange that may help us here.

Naram in Sumeria

Don't think having a giant as your leader was strange during this time. For consistency, let's look at a similar tablet from a similar time. This time the palette is from Sumeria and the ruler is named Naram-Sin. I know that sounds suspiciously close to Narmar, but forget about that right now. Again, we see that this leader is huge, his soldiers have long heads, but this time, Naram is the one wearing the helmet of horns that all Anak or Annunaki rulers wore in Sumeria. They, like the other Narmar, would have been worshipped as a god. Some would suggest the first Egyptian ruler was of the Anak race and he could possibly have even been this guy.

There are what appear to be 2 suns or some machines flying overhead. Naram's weapons are strange as well, while he is holding a spear thing the back looks very strange. He also has something that looks like a hatchet, but the head has 2 rivets purposely engraved on the tablet they were so important. His army has long heads as well. Both Narmar and Naram are bearded and wear the same type of Kilt.

I'm not getting into who this Naram-sin was, but one additional characteristic to consider is the flying fish. Sargon the king of Sumeria [possibly this same guy] is depicted in the next image as having a flying "thing" overhead and even with Sargon flying the thing. I know it doesn't look like a fish so there cannot be a connection of Naram and Narmer, but I'm still believing they could be the same giant Anak ruler either wearing horns or having others wear them, having long headed armies that destroy all comers, that has strange weaponry possibly to include flying ships of some type.

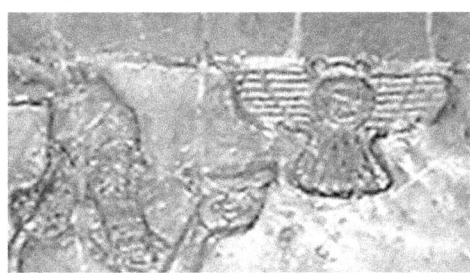

Sumerian Dinosaurs

In Sumeria they also found a seal and guess what was on it? The very same dinosaurs that were found on the Egyptian Narmar tablet. If you listen to an Egyptologist he will tell you that the dinosaurs represented the upper and lower parts of Egypt. Of course that is hog wash. Notice that the faces are even the same; so let's look at those horns.

Sumerian Horns

Just like on the Egyptian Narmar tablet, all the ruling Sumerian "gods" of the day had horns. This horn thing doesn't seem to be an Egyptian carryover, but in the sky with the flying fish things are these massive horned people "protecting" Narmar, so the Egyptian Tablet is showing

Sumerian Deities protecting him. Just a few of the hundreds are depicted. The collage below is a sampling.

So What?

I know you are thinking so what and what does a flying ship, dinosaurs, pulling a nose, giant long heads, and horns have to do with races of men? What if the long headed Anak ruled in Sumeria and Egypt, Ham and his group came along and settled on the side away from the Anak people. They promised not to bother them, but soon there was some sexual exchange of DNA modification that made a hybrid. Oh! I forgot to tell you Meritanten married a Scythian named Mille who would have carried the R1 mutation and she might have a few others which could have been the beginning of what is known as Rib. For this we need a little more history as Mille was king of Spain and the Gauls. Now for the news to find the origin of a mutation, simply look for the highest concentration of that mutation. Here is what we find.

Catalonia, Spain -82.5% have the R1b mutation; Ireland-82%; Britannia 80%; Normandy 76%; Scotland 72%; England 67%. From there the percentages drop quickly, the following map shows where the R1b is today.

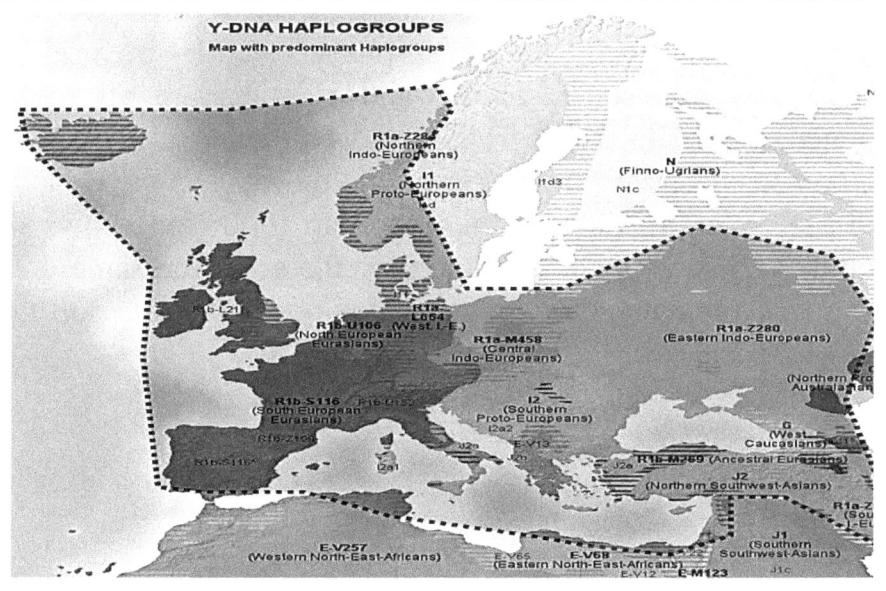

Notice how the top of Egypt has this unusual mutation?? If someone carried the mutation to modify the R1 haplotype to Spain and then to Ireland we would find most people there mutated. If Akhenaton's daughter carried the mutation, looked like the father and married the Spanish king before conquering Ireland, one might believe the Pharaohs with this unusual abnormality were Anak. That is not saying the people of Africa were or are of the Anak race. We can quickly see a difference. For one thing they are not red headed, or have a strong chin, or are Caucasian. They are different.

Hamitic Race

So we have old Ham wandering around Africa and making Hamites. As I mentioned Ham [Haplotype J or J1] and his group invaded Africa, but stayed pretty much by themselves. Soon there was some mixing and Hamites are all darker skinned, similar to the negroid race. That being said, there are differences. Hamites are characterized by some as the Bedouin and the covered a large area. The map following shows the Hamite people of today. In the map the areas the same color as Saudi Arabia are all Hamites. They now cover the northern part of Africa.

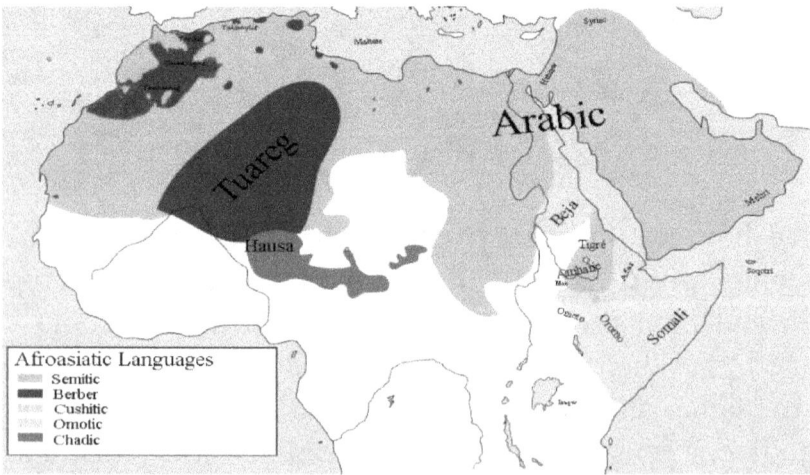

In the 19th century, European authors classified the Hamitic race as a sub-group of the Caucasian race, along with the Semitic race – thus grouping the non-Semitic populations native to North Africa, the Horn of Africa and South Arabia, including the Ancient Egyptians as Hamites. While very dark skinned, there was no mistaking Hamites for those of the Negroid race. Here are a few of their characteristics.

Forehead: *is high and square instead of low and receding*

Nose *is narrow, with the nostrils straight and not transverse*

Chin: *is small and slightly pointed instead of massive and protruding*

Hair: *is long and not woolly*

Lips: *are thinner than those of the negro and not everted*

Here are a few of the Hamite Race of people from northern Africa and Saudi Arabia.

That brings us to the Negroid Race.

Negroid Race

There is no doubt that the Negroid race is a noble one. Some of the images are shown below of this race of people from Nigeria, Sudan, and the Bantu. The race is still almost pure.

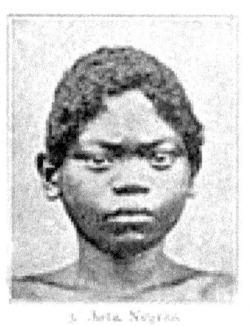

3. Jola Negress

1. Hausa, Western Sudanese Negro

2. Zulu, Bantu Negroid

Instead of just saying people someone looks a certain way they came up with a fancy name -- craniofacial anthropometry. Here is what is said about the Negroid race. The race is known by their broad and round nasal cavity; no dam or nasal sill; Quonset hut-shaped nasal bones; notable facial projection in the jaw and mouth area; a rectangular-shaped palate; a square or rectangular eye orbit shape; and large teeth. So here is what we have.

Nose: *flattish nose, flat root of the nose,*

Ears: *narrower ears*

Skulls: *narrower joints, frontal skull eminences, later closure of premaxillary sutures*

Hair: *less hairy, longer eyelashes*

Teeth: *cruciform pattern of second and third molars."*

Brain: *Slightly larger than Caucasian races*

Hair texture: *tightly coiled, kinky hair. It is a ubiquitous trait among Negroid populations*

Skin pigmentation: *varies from very dark brown to light brown*

Haplotype E

The following map of Haplotype E [African DNA] shows, Egypt is especially void of intermarriage and influx of African bloodlines. By staying isolated, their race stayed separate.

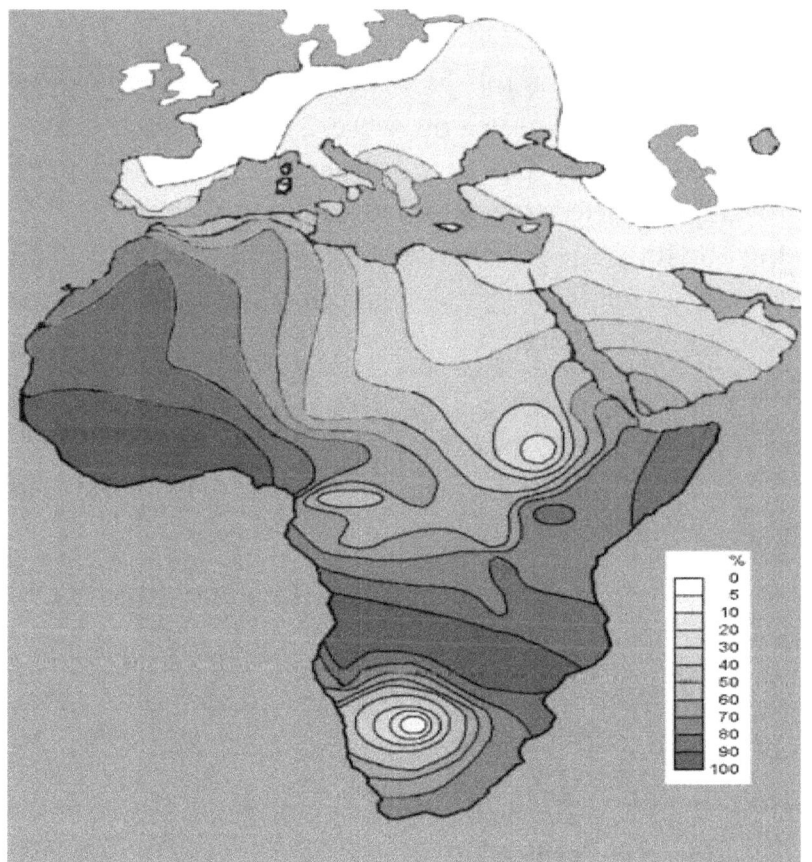

Conclusions

Hopefully, you can take something away from this book and enjoyed reading it as much as I enjoyed researching it. I know I have presented some odd things, but no matter how you feel about the strangeness, you must admit it gets rid of a number of the anomalies you were "Taught" in school.

- The Homo-Gigantus people who lived with the dinosaurs were very strange. I know you were taught over and over again not to read the Greek mythology unless you were ready to laugh as the stories had no meaning, but they told you a lie. Reading the 6th chapter of Genesis confirms their stature and Genesis one tells about their preexistence before the heaven wars.

- You don't want to believe in a "Heaven War" so the idea of the Anak people escaped you. Without the Anak, no genetic manipulation that was written about in dozens of ancient texts could possibly be true.

- Everyone told you, life came out of Africa and you bought into it hook, line, and sinker.

- Adam/Cro-Magnon was a real person and his clan was a significant change in humans.

- The Anak people, while different that use in some ways were still people and they still added "alien DNA" that was useful in our development.

- You must understand that the timing as described by nuclear decay have been expanded due to variation sin nuclear decay itself. It is still supported by the staunch evolutionists, but most are now have a horrible time as tests at multiple sites provide differences in dating by millions of years as nuclear decay is NOT REGULATED.

- You must understand that after tens of thousands of years of civilization, man was VERY civilized before the great world war. Sure no one told you and you thought everyone was throwing stones at each other, but the evidence doesn't support it.

- You must understand that there was a worldwide flood. Certainly not everyone died, but many, many died as a result. The earth DID change its rotation 10 thousand years ago and something like the moon of Venus exploding DID send ½ million meteors down on the earth.

- You must understand that chimpanzees evolving or being designed by humans rather than the other way around is easier to do and fits into the DNA mutation seen.

- You must understand that ANAK were giant long headed, red skinned, massive brain, long living people and around the world. While not gods, they were worshipped as gods.

- Around the world, people began mimicking the Anak to be like them. Some painted themselves red; others squished their head or wore long helmets.

- Ten thousand years ago, the earth shifted on its axis and many mutations occurred. This also caused many to die, massive flooding, and quick frozen mammoths.

- Six thousand years ago a massive war started and it didn't end until the earth was in shambles, many people had died, cities were in ruin, many went underground to live, and nuclear weapons peppered the sky causing massive mutations.

- If we can believe long heads, it seems the ancient Pharaohs were related to the Anak race.

- One of the Pharaoh daughters decided to leave with a Spaniard and conquer Ireland.

- Hamites took control of the Northern part of Africa while the E mutation people stayed towards the lower portion of Africa making the Negroid race.

- There is a possibility that one of the major mutations was that some people had apes instead of normal children for a time. This was not sustained and ape-people disappeared over time.

- I won't mention lizard-like people because it just makes me sad.

- I won't mention Ape- people for the same reason.

- Haplotyping studies have told us a lot, but one thing we must understand. Much of it still doesn't make sense. We must think beyond what people tell us in school or in books designed to insure no one jeopardized some sacred history or scientific theory.

While I didn't want to simply repeat the details of the book, I wanted to reemphasize that we are who we are because of those who lived before us. We cannot ignore history, religion, science, artifacts, and commonsense to build a mathematical characterization of DNA, without first understanding something about the races of man.

The End

About The Author

Steve Preston is a long lime author of scientific, esoteric facts. His books focus on the painful truths rather than whitewashed details that make us comfortable. If you are interested in the truth instead of comfort, please review other works by Mr. Preston as shown below. The images are some from Egypt taking the older version of taxi and in the Negev desert of Israel where the Dead Sea Scrolls were found and near where the beginnings of the Cro Magnon race began.

Searching at Egyptian Pyramid   Searching in Israeli Negev

www.ingramcontent.com/pod-product-compliance
Lightning Source LLC
Chambersburg PA
CBHW051905170526
45168CB00001B/250